Contributions To Phenomenology

In Cooperation with The Center
for Advanced Research in Phenomenology

Volume 93

Scope

The purpose of the series is to serve as a vehicle for the pursuit of phenomenological research across a broad spectrum, including cross-over developments with other fields of inquiry such as the social sciences and cognitive science. Since its establishment in 1987, *Contributions to Phenomenology* has published more than 80 titles on diverse themes of phenomenological philosophy. In addition to welcoming monographs and collections of papers in established areas of scholarship, the series encourages original work in phenomenology. The breadth and depth of the Series reflects the rich and varied significance of phenomenological thinking for seminal questions of human inquiry as well as the increasingly international reach of phenomenological research.

The series is published in cooperation with The Center for Advanced Research in Phenomenology.

More information about this series at http://www.springer.com/series/5811

Frank Schalow

Toward a Phenomenology of Addiction: Embodiment, Technology, Transcendence

Springer

Frank Schalow
Department of Philosophy
University of New Orleans
New Orleans, LA, USA

ISSN 0923-9545 ISSN 2215-1915 (electronic)
Contributions To Phenomenology
ISBN 978-3-319-66941-0 ISBN 978-3-319-66942-7 (eBook)
DOI 10.1007/978-3-319-66942-7

Library of Congress Control Number: 2017950519

Printed on acid-free paper

This Springer imprint is published by Springer Nature
The registered company is Springer International Publishing AG
The registered company address is: Gewerbestrasse 11, 6330 Cham, Switzerland

For us everything is concentrated on the spiritual,
we have become poor in order to become rich.

Friedrich Hölderlin

Preface

The topic of addiction has inspired countless number of books, with more being published everyday. In an area dominated by behavioral psychology, sociology, and medicine, however, there have been few attempts to tackle this vexing problem from a philosophical perspective. The study undertaken here ventures into unfamiliar territory, by transposing within the wider compass of the human predicament a problem, which the health-care community approaches primarily in symptomatic terms. To accomplish this end, I will draw upon the insights of key thinkers from the philosophical tradition of phenomenology and hermeneutics (e.g., Martin Heidegger, Maurice Merleau-Ponty, and Paul Ricoeur), who speak both directly and indirectly to the problem of addiction that begins to reach epidemic proportions toward the end of the twentieth century (and beyond). By employing the insights of these thinkers to reinterpret and understand anew the individual's struggle with addiction, the philosophical approach of phenomenology ceases to be confined merely to an academic discipline. Instead, such an approach has an interdisciplinary appeal, that is, by reopening a dialogue with the adjacent areas of transpersonal psychology, theology, and literature, in order to cultivate a broader perspective required to address the *changing landscape of the problem of addiction.*

In the last decade, this landscape has shifted dramatically due to rapid advances in technology, particularly as a result of the global outreach of the Internet. Almost every week, a new acronym arises to describe a potentially addictive offshoot of our use of cyber technology, the most recent of which is "shopping under the influence": that is, consuming alcohol while employing various "apps" to purchase products from one's computer, tablet, or phone. This latest form of compulsive behavior seems to combine two different kinds of addiction, e.g., "binge" drinking and impulsive shopping, into a third or hybrid possibility by which the individual can further relinquish his/her inhibitions. As facetious as this example may be, the Internet has nevertheless opened up a "virtual" Pandora's box of ways to become captivated by one's impulses and ultimately fall prey to addiction. Today, addiction has ceased to be an isolated problem reserved to an unfortunate set of individuals. Instead, technology has spearheaded the transformation in which addiction to this or that specific substance, e.g., drugs or alcohol, has given way to a completely new digital venue, transfixing the individual by the allure of a *virtual world of instant gratification*. The Internet not only becomes a conduit for various ways to become addicted but also gives birth to a completely new category: that is, "cyber addiction," the compulsive use of digital technology, which becomes dangerously apparent, for example, when people "text" and drive at the same time. With the Internet becoming universally available, and a culture of immediate gratification reinforcing our impulses, it should not be surprising that the problem of addiction has escalated in recent years.

Given these preliminary remarks, let me outline the central themes comprising my inquiry into this rapidly changing *phenomenon*. In Chap. 1, I will highlight the importance of developing a philosophical perspective to address the problem addiction, in contrast to other approaches in the fields of psychology, sociology, and medicine. In the second chapter, I will show how the "life situation" of the addict can provide an example or, in Heidegger's terms, "formally indicate" the pre-philosophical, pre-theoretical background that we all share—addict and non-addict alike—out of which any genuine understanding of the problem of addiction can first arise. Following the basic precepts of Heidegger's hermeneutic phenomenology, we will appeal to the life story or testimonial of the addict, in order to make explicit the individual's immersion in the struggle of existing, the foremost issue of having "to be," as the pre-theoretical backdrop for understanding addiction. In this way, a broader philosophical horizon will emerge that will enable us to address the phenomenon of addiction outside the traditional dualisms of mind and body, spirit and nature, which underlie psychology. In the third chapter, I will outline the existential roots of addiction; these originate from the structures of everydayness and harbor the *possibility* of an individual's becoming victimized by the pursuit of his/her self-indulgences. In the process, we will describe how our simplest desires can be exaggerated into "fetishes" and thus cross over to form the "hook" of addiction. How the manner in which the tendency to disguise, hide, and, most of all, "deny" the problem of addiction determines this "disease" constitutes the topic of Chap. 4. We will discover that these deceptive practices form a web entangling the addict and "enablers" alike, thereby masking the problem *as* a problem, on the one hand, and fostering a climate of *codependency*, on the other.

In Chap. 5, I will examine the birth of the "addiction crisis," as it broadens its reach within the technological culture of the Internet and its global distribution of the "means" for more people to become addicted. The question then becomes whether the global situation spearheading addiction today exceeds in complexity the insights of any single scientifically based narrative. To develop this critical perspective, I will outline the nexus of variables that create a climate for addiction(s) to flourish, that is, the "perfect storm" in which our inherent frailties as human beings intersect with an extensive culture of "enabling." While the "human-all-too-human" constants of the problem remain, our understanding of it must evolve in order to recognize how our *culture of conspicuous consumption* and its "commercialization" of fetishes extend the "hook" of addiction from all corners of the globe. Indeed, only by first confronting the wider scope of the addiction crisis, and the technological influences that intensify its grip on humanity, can we address *a new existential challenge*: i.e., that the rise of the "artifice" masks our capacity for self-understanding, even while extending access to both "information" and "entertainment."

In Chap. 6, I will consider the crisscrossing of various intellectual pathways that led to developing the first program to treat addiction (i.e., the so-called Twelve Steps), thereby outlining the *historical-cultural backdrop* or *interpretive horizon* within which its founder attempted to understand the problem. We will discover that the historical account of this development indirectly reveals a *gap* in the program for treating addiction, which *is only partially closed by appealing to religious maxims*: specifically, the paradox of how a physically based pathology can be overcome through the practice of a "spiritual discipline." In Chap. 7, I will show how this gap can be bridged through a therapeutic approach that is informed by a philosophical understanding of the self through the temporal-spatial trajectory of its *embodiment* and its tendency to become entangled in the deceptive practices from which a vulnerability to addiction arises. Rather than objectified by the natural sciences, the self reemerges through its immersion in the human predicament, the crisis that it spawns (including addiction), and the struggle to cultivate new horizons of meaning, e.g., through the capacity for transcendence. As we develop our understanding of addiction from out of the individual's concrete life experiences, the language by which we address this *phenomenon* (of the tendency to become addicted) also changes.

In Chap. 8, we will show how our attempt to carve out a new landscape for understanding (and, indeed, speaking of addiction) implies a strategy for treating the illness. Specifically, any process of recovery or healing must begin from the individual's capacity for self-understanding and the transformation power of the moment (*Augenblick*) to offset the fetish-like pull of immediate gratification. This recovery and healing must be conceived anew as a path recoiling upon itself in an elliptical orbit that returns, as it were, "each day" to anchor the individual in the "gifting" and initiative of freedom, of "choosing to choose" again and again. The self-inducement of this transformative way of temporalizing departs from the linearly based, clinically oriented model that is assumed as the theoretical premise of various treatment programs. The hermeneutic-phenomenological method redirects the individual to a path whose coordinates are defined as much by the challenge of discovering the meaning of life, as in submitting to any single model of treatment. In this way, the *path* to recovery lies in challenging each individual to reexamine his/her set of priorities and the freedom on which they are based. The challenge of rediscovering the self's life trajectory, of its striving for transcendence, opens this pathway of healing and recovery.

In the clinical search to find a "cure" for addiction, we return to the basic presupposition concerning the power that language has in prompting the self to take ownership for its existence. Even in the midst of the latest medical advances, the *phenomenon* of addiction may instead remind us of the frailties of human life and how our triumphs are always mired in the conflicts we overcome. When viewed within a larger historical context, addiction is not just about the plight of the addict, but instead is also a signpost in the search for a meditative place of stillness otherwise lacking in our fast-paced culture of conspicuous consumption.

New Orleans, LA, USA Frank Schalow

Acknowledgments

There are more people than I am able to thank who have contributed to helping to shape the development of this book. First, I wish to express my gratitude to Kenneth Kahn of the University of New Orleans, for planting the seeds of the idea for this project. I also credit two of my early mentors, Francis F. Seeburger of the University of Denver and Michael E. Zimmerman of Tulane University/University of Colorado at Boulder, for paving the way to apply Martin Heidegger's phenomenology for the study of addiction. Parvis Emad and George Kovacs have provided valuable advice and encouragement throughout the completion of this book, along with M. J. Hannush, the former book review editor of the *Journal of Phenomenological Psychology*. Among my colleagues at the University of New Orleans (too numerous to name), I would like to thank Robert Dupont, chair of the History and Philosophy Department, and Mostofa Sarwar, director of the Honors Program and associate provost, for their continuing support of my research.

I would also like to thank Cristina dos Santos and Anita Fei van der Linden-Rachmat of Springer Publishers, along with the editors of the "Contributions to Phenomenology Series," Nicolas de Warren and Dermot Moran, for their supervision in guiding this project through the review and production process, as well as Emily Orfanides for her help with proofreading the manuscript.

I graciously acknowledge the editors of the following publications, for granting me permission to use portions of these articles in my book:

"Fantasies and Fetishes: The Erotic Imagination and the Problem of Embodiment," *Journal of the British Society for Phenomenology*, 40/1 (Jan. 2009): 68–82, ed. Ullrich Haase.

"The Phenomenological Elements of Addiction: A Heideggerian Perspective," in *Horizons of Authenticity in Phenomenology, Existentialism, and Moral Psychology: Essays in Honor of Charles Guignon*: 165–178. Edited by Hans Pedersen and Megan Altman (Dordrecht: Springer, 2015).

Contents

Part I
The Origins of the Addiction Crisis

Chapter 1
Seeking a Philosophical Perspective

1.1 Preliminary Considerations

For thousands of years, philosophy has taken its point of departure from what is most question-worthy, from the wonder provoked by seemingly the simplest and most elemental concerns. In what is apparently obvious, the philosopher discerns what is most odd, unique, and even provocative. Even today as science promises more (and perhaps "faster") solutions to life's problems, we are beset, if only occasionally, by the enigmas, riddles, and even paradoxes of the human condition. Where the opportunities for entertainment and amusement are to be found no further than the touch of one's cell-phone, boredom, malaise, and depression still abound. Indeed, as the struggle for mere survival and the fulfillment of life's necessities diminishes, so does our vulnerability to affects and emotions increase in almost inverse proportion. In wake of these changes, traditional philosophical concerns as to what it means to exist, the human predicament, along with our sense of embodiment as natural creatures, become harder to pinpoint and define. Now all that pertains to the use of our hands, from gestures to actions, becomes pressed into service through an *artificial appendage*, that is, the hand-held device. As the separation between the tool and its user begins to blur, so does the potential arise for displacing the fields of concerns and relationships that ordinarily "situated" and oriented human existence (and its search for self-identity) into the allure and attraction of this hyper-stimulated domain of images that we call "cyberspace."

In this overwhelming rush of sensory stimuli, the age-old problem of the individual's darkest tendency to succumb to self-destructiveness intersects with the newest, glittering glamor of technological wizardry. How do we characterize this intersection in simpler terms, which reflects both the recurrent problem of our human frailties, on the one hand, and the most recent change in the *modus operandi* of our being-in-the-world, on the other? In a word, we can point to the pervasive and ubiquitous phenomenon of addiction to describe this new configuration of the

F. Schalow, *Toward a Phenomenology of Addiction: Embodiment, Technology, Transcendence*, Contributions To Phenomenology 93,
DOI 10.1007/978-3-319-66942-7_1

pathos of the human predicament. In making this observation, however, our way of *speaking, describing, and alluding* to this phenomenon must change in concert with developing a deeper and richer understanding thereof. Because the *nuances of language* become central, a philosophical approach must take precedence (over traditional ways of addressing addiction). For by the word "addiction" we no longer simply mean a clinical or medical diagnosis concerning a person's dependence on a specific substance, e.g., drugs or alcohol, but instead designate a broader historical and cultural transformation of our way "to be" (as well as an individually-based problem).

Due to the pervasiveness and closeness of our familiarity to it, addiction may very well qualify as an issue that merits philosophical scrutiny. Traditionally, the study of addiction has been reserved to the field of psychology and its primarily clinical mode of discourse. Of course, understanding addiction is one thing, and developing an efficient plan of treatment is quite another. Yet, if philosophy could contribute to such understanding, then it may not be simply due to the insightful ponderings of philosophers, as much as the increasingly provocative character of the concern, issue, or even phenomenon under scrutiny. Could it be the case that addiction is a harbinger of shifting cultural horizons, and is as much emblematic of an impending (human) crisis as a whole, rather than only a medical or clinical problem? By providing an entry point to this historical crossroads, perhaps the issue of addiction can become question-worthy in its own right, that is, *turn into a philosophical problem.*

Could it be that the renewed concern for addiction, the need to wrestle with its recalcitrance and revisit it *as* a problem, directs us back to the perennial philosophical questions concerning what it means to be human, to exist, and, ultimately, the meaning of being in general? In addressing this question, two observations emerge at the outset, which will also point us in the direction of developing a methodology unique to our investigation. First, if addiction is not necessarily unique to our time, then its ubiquity may very well be, and, indeed, is in part due how our *Zeitgeist* privileges the instantaneousness of the 'moment,' of the raw instant, through the optics of our technological gadgets; if addiction activates a basic human urge to seek "immediate gratification," then there has never been a greater opportunity since the birth of civilization to deploy, literally, various platforms or mechanisms of self-indulgence. Second, the individual who embarks upon the path of recovering from addiction must still confront the basic challenge of what it means to exist, the bare fact of the struggle "to be" endemic to all of us—regardless of how the psychoanalytic, medical, and clinical community adjusts its diagnosis of the "disease" of addiction and pioneers new programs of treatment, e.g., through applications in neuroscience and neurochemistry.

Philosophy exemplifies a journey in which the questions posed require reformulation, in an effort to broaden and expand their scope and relevance. In this regard, the preliminary concern, hesitation, and pause as to what philosophy may contribute to a topic otherwise monopolized by psychology changes into the question as to what is the source of interest that could attract, evoke, and invite an appropriate level of philosophical scrutiny. In light of such a philosophical orientation, as we will see, the concern for the overall approach or methodology of our study takes precedence. From a psychological vantage point, for example, the tendency might be to address (the issue of) addiction by isolating and abstracting out a cluster of characteristics, e.g., neurological components and behavioral aspects such as obsessive-compulsive tendencies, which might suggest a specific diagnosis for the aberration. Instead, my aim is to address addiction as a problem surfacing within the concrete life-situation of the individual, in such a way that the vulnerability to and ongoing recovery from addictive behavior emerge as two sides or possibilities of the self's way "to be," e.g., of confronting the basic challenges and quandaries of human existence.

Thus, while addiction may be seen as beginning from physiological cravings, we discover that these addictive tendencies (e.g., experienced *as* cravings) spiral out from the wider compass of the body *as* lived, that is, as contextualized through a set of relationships and concerns, emotional responses, even "stress," which form the concrete tapestry of our being-in-the-world. In this regard, we must pay special attention to the methodology by which we can attend to the complexity of its phenomenal and experiential expressions, which speak not only symptomatically, but also *existentially* in terms of a larger life-predicament and search for meaning (in which the struggle and *pathos* of our lives are interwoven).[1]

Addiction has been called the "universal plight" or "epidemic" of our time. We must differentiate, however, between addiction as a medical issue (which goes back many centuries) and its emergence as a culturally pervasive, perhaps even historically induced *crisis* in its own right. As Ryan Kemp states: "The levels of addiction have increased over the past two hundred years. There is no coincidence that this has happened at the same time that modern industrial based city-society has developed."[2] In the United States, particularly, the emergence of addiction as a ubiquitous and vexing problem stretches across generations. Throughout the past century, one constant remains true: whatever approach we may employ, religious or medical, moral or psychological, a stubbornness continues to persist in our struggle to come to terms with addiction *as* a problem, much less develop an effective program for its treatment.

In this regard, I will develop a philosophical approach to study addiction which, while it welcomes the insights of the psychological and clinical community, in its basic orientation stands in stark contrast the narrower vision of the so-called "scientific" (or even medical) establishment. How are we to identify this approach with a distinctive methodology originating within the Western tradition? Although clues may already be apparent, we defer momentarily from applying any specific label, in order to a broader direction of questioning that can emerge (than can be confined to any one thinker or even philosophical school). Indeed, the eclectic character of my approach *precludes endorsing the vision of any single thinker* or even philosophical movement, although, we will see, specific figures will come to the forefront insofar as their insight will resonate with, and be emblematic of, the *Zeitgeist* or "spirit of

[1] For a discussion of how the method of hermeneutic phenomenology impacts contemporary psychoanalysis, see Stolorow 2013, pp. 451–453.

[2] Kemp 2009, pp. 1–18 (esp. p. 16).

the times." The phenomenon of addiction is of such a multifaceted character as to dictate this eclecticism, as embodying the cultural evolution of a problem (and its proposed plan of treatment) and as a microcosm or mirror image of the larger human predicament.

1.2 Methodological Concerns

The philosophical, literary, and even psychoanalytic sources from which I will draw display a common ground. The key figures of this period share a *Zeitgeist* from which the development of the first program for treating addiction *as* a problem arises (e.g., in the 1930s); in unison, they thereby point to a common universe of discourse within which the problem of addiction can be addressed *as* a problem. Throughout this study, I will trace the confluence of philosophical, religious, and literary traditions out of which the first attempts are made to uncover the problem of addiction both in a clinical setting and as emblematic of the basic human predicament encompassing all of us. We will witness the coalescing of an intellectual tradition, which forms the historical backdrop for *understanding* addiction, insofar as it first emerges as a problem in the twentieth century and evolves into a cultural phenomenon today.

In employing the aforementioned term to *describe* the pervasive scope of addiction today, my approach is "phenomenological" in the most basic and broadest sense, but also in its distinctly *philosophical incarnation* of drawing upon the insights of historical figures who comprise a tradition and adhere to the simple rendition of its guiding principle: the return to the concrete dimension of lived-experience. Given this orientation, I will address addiction in terms of a unique synergy between outlining the human experience of it and developing a self-understanding of its basis in human existence. Proceeding in this phenomenology manner, I will then to take as my point of departure the simplest observance of a specific datum, i.e., a readily identifiable and obvious phenomenon or what we might in the vernacular describes as a *brute fact*. And what is the phenomenon in question? We wish to point simply to the basic conundrum of addiction, namely, the attempt to develop a rational, coherent plan or program to treat addiction in the face of the addict's incorrigibly irrational, e.g., "obsessively-compulsive" bent to resist such treatment in behalf of succumbing to whatever it is he/she may be addicted.

Because the phenomenon in question is multifaceted, and exhibits many different permutations, our inquiry is also "interpretive" at its core. The task of interpretation is buttressed by the fact that the issue of meaning in the largest sense permeates the problem of addiction. This permeation occurs both on the philosophical side of the attempt to address the essential elements of the human predicament and the personal side of the addict's need to cultivate his/her own self-understanding as a path to recovery. The fusion of these two horizons of meaning,[3] as it were, yields a

[3] See Gadamer 1967, p. 39.

historical-cultural platform on which to conjoin the different perspectives of the addict and non-addict alike (not to mention, that of the therapist and patient). The unfolding and interpretation of this historical-cultural backdrop belongs to the task of what we can broadly construe as hermeneutics.

As I sharpen and define my phenomenological approach, I will point to a litany of figures whose insights are crucial to forging a new understanding of this topic. The *Zeitgeist* that directs us to the existential roots of addiction, however, also includes a critical perspective to address the impact that technology has in creating new opportunities for addiction and facilitating *instantaneous* access to them. In this case, the attempt to outline the dynamism of time in the concrete unity of its existential, cultural, and historical dimensions defines an important step in the development of the phenomenological tradition. While the prominence of any one figure can be debated in the scholarship, his/her importance to this study depends upon providing an initial inroad into addressing addiction and thereby *initiating a discussion via key motifs etched in a temporally based vocabulary.*

Because in his *magnum opus*, *Being and Time* (1927), he explicitly addresses addiction as a phenomenon, I will take my point of departure from the figure who first integrated phenomenology with hermeneutics, namely, Martin Heidegger (1889–1976). In the 1920s, Heidegger pioneered a strategy for exploring the meaning of human existence, and, subsequently, extended his insights into temporality and history to address the complexities of the modern age of technology. Because the problem of addiction shapes the broader horizon of our contemporary *Zeitgeist*, the scope of my study, as indicated above, must broaden significantly from its focus on any one thinker. In order that the breadth of this horizon may gradually become visible, I will engage in a form of "context-building," in order to accommodate diverse perspectives and interweave them into a broader philosophical vision. As part of this context-building, I will outline the confluence of sources both backward and foreword: from intellectual influences on Heidegger, to his influence on the development of the founder of an approach to existential psychotherapy, Medard Boss,[4] to his apprenticeship under the Swiss psychiatrist, C. G. Jung (1875–1961), who first advised a patient struggling with addiction to seek a religious perspective, who in turn paved the way for developing the first program of addiction by his friend (and fellow addict), Bill Wilson, the primary author of the "Twelve Step Program." I do not intend to provide a historical account of various causal influences and chronology of events in the development of the first program for treating addiction (of which there are a plethora of books).[5] Instead, I will outline the

---

[4] Boss 1963, pp. 45ff.

[5] See Kurtz 1991, p. 33. Kurtz provides a historical chronicle of the founding of the first program for treating addiction, which includes among its key "'founding moments' in the history of the idea and the fellowship of Alcoholics Anonymous … Dr. Carl Gustav Jung's 1931 conversation with Rowland H." and "Ebby T.'s late November 1934 visit with Bill Wilson…." It was after this visit that Bill Wilson reported having a life-changing, "white light" experience in middle December, 1934. While traveling to Akron, Ohio (Spring, 1935) on a business trip, Bill Wilson forged a friendship with Dr. Bob Smith, himself battling alcoholism, and together they would become "co-founders" of AA.

genealogy of sources that interplay to shape the intellectual tradition both for understanding and treating addiction (from the 1930s until today). What this genealogy reveals is the unfolding, in Heidegger's words, of a philosophical insight or "moment of vision," or, in Jung's terms, a remarkable "synchronicity" in how various traditions converge to shine a common spotlight on a perennial problem which pervades our lives presently. Borrowing again from the European tradition, we can characterize this convergence of diverse sources of insight as displaying a *Zeitgeist* or *a shared intellectual vision of the time*.

We cannot discount the advances made in neuroscience in regard to treating addiction by regulating brain chemistry. But even so-called "drug-therapy," no matter how successful it may be in treating an array of physical-psychological disorders (including depression), cannot by itself solve the "riddle" of human existence. Among the first to identify the limitations of this behavioristic model for resolving human conflict was Aldous Huxley (1894–1963), who was prominent is his own literary field as the aforementioned luminaries were in their areas. In 1932 Huxley published one of the most influential novels of the twentieth century, *Brave New World*; he brought to light in a literary, satirical genre, the imminent threat of a "world" dominated by humanity's submission to technology.[6] While Huxley may only indirectly be included within the litany of the aforementioned genealogical sources (addressing the topic of addiction), he was every much a visionary as the others. Indeed, Huxley's satirical depiction of technology predates by 4 years Heidegger's breakthrough in describing technology as an encompassing, ubiquitous form of "machination," in his manuscript from the period 1936–1938, *Contributions to Philosophy*. In one of his most influential lectures delivered, "The Question Concerning Technology," Heidegger gave deeper expression to his claim that the modern age had become dangerously dependent upon its use of technical devices.[7] The issues raised by the subsequent, exponential growth of technology are not incidental to our study. While Heidegger lived into the computer age and witnessed the growth of cybernetics,[8] he did not foresee the rise of the Internet and its corresponding potential for extending the avenues of addiction (from online "gaming" to pornography) in our "digital age." If instant gratification is the chief *modus operandi* of addiction, then the Internet unleashes that force in ways that were unimaginable when the first program for treating the problem was created in 1933. How, then, does technology appear in our digital age? *Put simply, technology appears as the inconspicuous backdrop for the mediation of human desires on a global scale, assimilating them into a vortex of the self's pursuit of immediate gratification and the fulfillment of its "fetish-like" cravings, e.g., addictions.*

In philosophy generally, and in a philosophical approach to addiction specifically, we ask questions that might appear obvious, even unnecessary to others. Yet, raising such questions may also plant the seeds for shifting the framework, a

[6] Huxley 1968. *Brave New World* was originally published in 1932.

[7] Heidegger's understanding of technology came to fruition through a series of lectures delivered in 1949 at Bremen. See Heidegger 2012, pp. 23–43.

[8] Heidegger 2003, p. 63. Also see Heidegger 1989/1999, pp. 120–130; tr. 84–90.

complete paradigm shift, perhaps, for approaching such a perennial problem as addiction. Why, then, is addiction so pervasive? In asking what might appear to be obvious, we challenge the conventional wisdom that suggests addiction is primarily an "aberration," if only by exemplifying a reckless life-style. In seeking a new point of departure we our study, we must consider, however, whether the opposite may very well be the case. That is, on a deeper level, the tendency to become addicted may qualify as the "norm," namely, by pointing to a destabilizing force endemic to the human predicament such that each individual may be vulnerable to *losing* his/her sense of self-direction in life.

By applying the method of hermeneutic phenomenology, I will bring this problematic dimension of the human condition to light. In Heidegger's terms, I will begin from the juncture common to us all, a shared starting place (encompassing addict and non-addict alike), namely, our immersion in "everydayness."[9] By exploring how each of us, as instances of the *act of existing*, are also bound up in and delivered over to this "way of being" of everyday life, we will consider how a tendency to become addicted radiates outward from the individual's pursuit of his/her most commonplace concerns. In this way, we will introduce a new language to address the multifaceted character of addiction, in terms other than what are ordinarily used by clinicians who may inadvertently monopolize the discussion of this pervasive social problem. Despite many benefits, the clinical approach also flounders in its attempts to quantify and objectify aspects of the human condition that are beyond such quantifiable methods. Given this new linguistic bent, we will open up new vistas in the study of addiction, and consider nuances in its origin and development that remain otherwise concealed in many studies. Conversely, in addressing the problem of addiction in this way, we must also develop an interpretation of the constitutive elements of what it means to be human. What I have already alluded to as the human predicament will appear as situated in terms of our being-in-the-world, that is, through a spatial-temporal horizon of meaning within which we can understand the tendency to become addicted (from the concrete life-experiences, i.e., including first-person accounts, of the individual's confronting this problem).

The advantage that phenomenology has in developing its own terminology lies in its orientation to the concrete level of human experience, whereby common usages from everydayness can still provide clues to the phenomenon of addiction. In everydayness, we refer to addiction as a "scenario" in which the individual is "hooked" to whatever emerges as the primary source thereof. Here we employ a metaphor to describe the beginning of addiction, in which an allure, urge, or craving takes hold and takes over the individual's ability to control it. As an illness that can be clinically diagnosed, addiction entails a combination of physical and psychological reactions, (e.g., feelings of dependency). But what is the experiential basis of this enticement, excitement, and source of attraction for this "hooking" (as a verb) to occur? What source in everyday existence discharges this potential to be captivated, enthralled, and subdued, however momentarily, which forms the crucial juncture of our being-in-the-world where the self's focus on a single preoccupation

[9] See Schalow 2003, pp. 17–30.

begins to predominate to exclusion of all other concerns? At this critical juncture, as the "zero-point" of addiction, the rise of a specific allure simultaneously begins to erode the individual's capacity to choose (i.e., exercise restraint and self-control), so as to shift the axis of priorities in favor of pursuing the source of one's addiction. This catalytic moment ensnares and enslaves alike, abandoning the individual to the fickleness of his/her cravings. This life-altering moment spontaneously induces a fixation that completely consumes the individual, thereby setting the "hook" of addiction. How can we characterize this magnetic force, the irrepressible allure of this fixation?

By flying in the face of any rational counsel, the pull of the magnetic force of enticement clashes with the individual's sense of self-preservation; an apparent irrationality abounds in which the addict proceeds along a self-destructive path, possibly succumbing to the compulsiveness of his/her cravings. How do we characterize this "hooking?" The term "fetish" provides us with a unique phenomenological description, which originates from the inter-affective dimension of the self's *embodiment* as a primary way of being-in-the-world.

1.3 The Phenomenon of the Body and the "Hook" of Addiction

In developing a new perspective on this pervasive problem, it is necessary to reconcile our analyses with the assortment of facts that speak to the individual's concrete experience of addiction, from the perspective of the addict and non-addict alike. In this way, we answer the primordial call of all phenomenology (including Heidegger's) of remaining true to the phenomenon as such, or the "thing itself," as it becomes explicitly manifest. In heeding this phenomenological directive, the unique idioms that have become almost universal in the language by which we speak of addiction, and the distinctive spell that it casts, provide the initial clues for formulating the two previous questions. Whatever may be the confluence of conditions that creates an atmosphere for addiction to flourish, the one inescapable element is a vortex of desire within which an addict can be "hooked" in his/her attraction to any given source of addiction. While this description has been integrated into our cultural lexicon, the question remains open: how does one become "hooked," and why one person and not another?

The specific logistics that are involved in an individual's becoming addicted, then, pose something of a paradox. On the one hand, the overwhelming and irresistible "allure" of the source of addiction subverts the individual's capacity to exercise self-control. On the other hand, the subtlety of the hook lies in promising an empowerment, whether in the form of exhilaration or escape, which exceeds that otherwise granted through the range of possibilities available within the environment. How we can explain this paradox, and others like it that pervade both our understanding and treatment of addiction, constitutes the challenge of applying the phenomenological method to address this problem. While it is easy to dismiss the paradoxes as mere obstacles to greater insight, they can also point to or indicate the distinctive phenomenon of addiction, not simply as an academic puzzle, but instead as radiating the complexity of the human experience thereof. To be sure, many theoretical models tend to simplify the problem, serving the dual interests of understanding and clarity. In taking a fresh look at the problem, however, it is equally as important to appreciate its inherent complexity.

In the course of this investigation, then, it will not be the power of any one insight that directs us, but instead a path of development that allows us to reconstruct the problem of addiction. In embarking upon this path, we will reshuffle the priority of the factors normally believed to be central to addiction, of least to most important. Rather than becoming immersed in the various "symptoms" of addiction, it will be necessary instead to address what they are "signs" of, that is, the overall nexus of meaning in terms of which these various symptoms can be "interpreted" as signifying the drama of the individual's struggle with addiction. Where phenomenology addresses the "phenomenon" in question, Heidegger designated hermeneutics as the unique angle for discerning, deciphering, and, ultimately, for "interpreting" the meaning of the individual's experience within the context of the larger "life-situation" in which the drama of the human condition unfolds. By transposing the problem of addiction back into this concrete setting, we will be able to: (1) reconcile our analyses with the concreteness of the individual's own "life-experiences," thereby preserving the integrity of the phenomenological method and (2) recover the "existential" perspective of the self's search for meaning and for the possibilities distinguishing its uniqueness, (3) and re-animate the discussion and its language, not just with technical terms, jargon, or even the buzz words of pop-psychology, but with key motifs that portray the epoch struggle of the human predicament in the most enduring way (e.g., Greek tragedy), and, (4) identify the *body as the lived, meaning-giving locus where the "fetish-like" attachment of desire can first unfold* and distinguish the existential basis on which the dependency of addiction rests. In this way, we will be able to extricate ourselves from a "one-dimensional" account of addiction, as "X" versus "Y," as "nature" versus "nurture," as "addictive personality" versus "environmental influences." Albeit from the stance of pragmatism, Bruce Wilshire emphasizes the importance of our "body-selves," as the meaningful locus from which we (in modern society) can experience the possibility of recovery, along with the self-destructive tendencies of addiction.[10]

In re-opening the problem of addiction, we will consider it not simply is it appears in terms of its symptoms. Instead, we will address the problem of addiction, insofar as it harbors a larger *gestalt* of meaning. Given this *gestalt*, we can glean how the individual's desires can be fractured into conflicting ends, which give inordinate power to one's "impulses," only at the expense of removing the ability to control them. The fracturing of desire in this way, however, is not simply a mental process, but is also connected with *the lived-experience and self-understanding of the individual's embodiment*. Instead, a "pyramiding" of meaning occurs, in which

[10] Wilshire 2003, pp. 3–6. For a parallel discussion, that emphasizes the connection of addiction and failed coping mechanisms, see Nakken 1996, pp. 29–33.

what may originally appear as meaningful within a commonplace setting, becomes transposed on a higher plane to trigger additional associations and connotations; while parasitic on the conventional sense, these new connotations allow a "precedent" of meaning to arise which is open to further amplification. When this new precedent of meaning becomes the exclusive focus of the individual's concern, and shifts the direction of the desire toward what withdraws and eludes any promise of complete fulfillment, then something on the order of a *fetish* emerges. The fetish not only harbors an added significance to what is conventionally understood, but also, because of this further dimension of the elasticity of meaning, has the potential to ignite greater interest, spark arousal, and even in the end monopolize the desire.

The possibility that everything a person invests with meaning can be assimilated to a single concern, and become an exclusive preoccupation, pinpoints the "allure" of the "hook" predisposing anyone of us to addiction. While not interchangeable with, or reducible to, addiction, the commonplace occurrence of "fetishes" explain an elemental truth: that addiction is an indiscriminate plight to which everyone is vulnerable. But the greatest advantage of looking to fetishes as a *window into the complex phenomenon of addiction* lies in completely shifting the axis for the study thereof. Because fetishes suggest an obsession (of what frequents one's thoughts) and perhaps a compulsion as well (of acting-out), which is not originally connected to a physical craving, they suggest another avenue for exploring the roots of addiction. Indeed, fetishes illustrate how one's desires can be diverted through an ambiguity of signification, providing a scenario in which the individual not only loses control, but also becomes caught in a web of deception, deceit, and denial. Accordingly, the study of fetishes makes explicit how "cravings," originating physically or otherwise, can become attached to "false ideas" or "illusions," and, by the same token, suggests why "process," "behavioral," or "activity" addictions have just as much importance as their "substance" based counterparts, in addressing the complexity of the problem. Put another way, the lived-body provides the existential lens through which a new avenue of fetish-like attachment becomes visible, namely, the artificial platform built around the iconic power of digitized images. As we will discover, the *rise of the artifice will pose a further hermeneutic question about the abrupt diversion and displacement of our desires* and their impact on our capacity for self-understanding. In recent years, this displacement has been further enhanced by development of the "hand-held" device, the "smart-phone," which brings the allure of digital images to our fingertips.

This hermeneutic transposing of the problem of addiction has important implications for how to reformulate the two central enigmas surrounding this topic. As already indicated, this "paradigm shift" provides the clue as to how we can consider "process" ("behavioral") and "substance" addictions to occupy the same ground,[11] insofar as allure of the "hook" lies in supplanting every other "meaning" in life.[12] And, likewise, the broadening of the horizon to understand the problem of addiction

[11] See Carnes 1997, pp. 9–10, 150–151. Also see Seeburger 1993, pp. 52–54.

[12] As a case in point, we need only look at the rise of cases of addiction through the use of prescription drugs, specifically opiate derivatives.

provides a new basis for re-examining the derivation and implementation of the "Twelve Steps," insofar as this program of treatment was initially formulated in the1930s as a response to only one kind of disorder (and the "substance" genre at that). Accordingly, if we are to re-open the problem of addiction, it becomes necessary to address the confluence of sources from which the first pioneer, Bill Wilson, developed the Twelve Step Program, as well as ask what are the keys to "working" this program. Once again, the unique window that fetishes offer into addiction also holds up a mirror to see, in reverse, how the grip of certain false ideas and illusions might be broken by encouraging the addict to submit to a litany of alternative beliefs, e.g., "giving oneself over to a higher power" (Step One). Could it be the case that what may appear as only "incantations"[13]—the reiterating of one step after another—actually "empowers" the individual to break the captivating spell of the addiction? Indeed, this is a practical question that we will attempt to answer in the course of our study.

1.4 A Shared *Zeitgeist*

Viewed in the estimation of some experts as an "epidemic" of our age,[14] the problem of addiction has proven to be as stubborn as any medical conundrum in the past century, as far as making inroads into developing preventive measures and better programs of treatment. Whatever cultural explanations we may offer, perhaps none is more obvious than the simple contention that the potential to become addicted is as inevitable and inescapable as any tragic situation in which human beings occasionally find themselves. The simple fact that new forms of addiction arise almost every day, and, whether under the label "sexual" or "shopping,"[15] "Internet" or "gambling," and, most recently, "video games,"[16] become part of the cultural lexicon, gives us occasion to pause. Indeed, the problem is getting worse and not better,[17] and its tragic consequences more poignant, not less so. And yet, while the plethora of books on this topic continues to accumulate, from scientific treatises to pop-psychology, perhaps a sufficient impasse has been reached to warrant addressing the problem of addiction from the vantage point of a completely different methodology and orientation.

---

[13] Heidegger 1977b/1962, pp. 259–260; tr. 240.

[14] See Pinsky 2003, p.159.

[15] See Seeburger 1993, pp. 34–35. This book, however, was written *before the emergence of the Internet as providing a broader platform for multiple kinds of addictions*. On the opposite (i.e., "medical") side of the spectrum, see Erickson 2007, pp. 153–155.

[16] Dr. David Greenfield can be credited for identifying this specific genre of addiction, and developing a program of treatment, in his newly founded "Center for Internet and Video Game Addiction." For a popular discussion, see Kardaras 2016, pp. 11–12.

[17] See Schalow 2006, pp. 20–34.

To be sure, the problem of addiction may be too multifaceted to yield to any single approach. My aim in this investigation, however, is to explore the depths of this problem by bringing it under the scrutiny of one of the most radical philosophical methods developed in the twentieth century, namely, hermeneutic phenomenology. In pioneering this method, Heidegger not only provides a critical counter focus to the "positive" or "objective" sciences[18]; he also supplies many tantalizing clues to addressing addiction,[19] as well as an overall critique of the technological culture in which various addictive behaviors flourish, so as to offer a worthy point of departure for re-examining this vexing problem. But in addition to what he explicitly says, or even implies, concerning the phenomenon of addiction, there is perhaps an equally important, if only a historical coincidence to be noted, which makes his discussion of addiction all the more compelling. Specifically, the twists and turns of Heidegger's own philosophical development interweave with the path along which his contemporary, Bill Wilson, traveled in piecing together the various sources to devise the first program for the treatment of addiction, the Twelve Step Program of AA (Alcoholic Anonymous). While each in his own right is as dissimilar as the other, both Martin Heidegger and Bill Wilson overlap despite their differences, insofar as the phenomenological methodology pioneered by the former can be implemented to address the assumptions on which the latter based his "program" for treating the problem of addiction. First, Heidegger was more acutely aware of his own intellectual roots than any of his contemporaries. Secondly, his method provides the keys to unravel the basic assumptions about *who* we are as human beings, a "pre-understanding," which forms the backdrop of any attempt to confront a distinctive problem of our humanity, including addiction. Thirdly, his understanding of the phenomenon of human existence centers on precisely what is *not* transparent, that is, the *individual's tendency toward dissimulation, distortion, and (self-) concealment as the hallmark of addiction*, i.e., of its pattern of *denial*. Heidegger, then, not only provides a critical perspective from which to re-evaluate the major intellectual movements in the development of the Twelve Step Program. But he also supplies the keys to look forward and ask: why has our understanding of addiction lagged behind its rapid mutation into the global problem it has become today?

1.5 The Interface between Religion and Therapy: Seeking a Program of Treatment

Our attempt to explore the origin and roots of addiction brings to light an important historical anomaly in the development of the first program of treatment. Indeed, this first program (what we all know today simply as the "Twelve Steps") emerges in a vacuum, not only for understanding addiction, but also how its implicit religious

---

[18] For further discussion, see Aho 2009, pp. 118–119.

[19] For an interesting way of raising this question, from a Hegelian standpoint, see Morris 2001, pp. 409–412. I am grateful to Professor John Russon for providing me with this reference.

orientation can *double* as a *de facto* form of therapy. The "Twelve Steps" program works, to the degree that it does, by importing a religious message, albeit without requiring a commitment to a specific religion (or even a supreme being, for that matter). According to the established historical account or narrative, the program of treatment imports a form of universalistic Christianity, which was adopted from the Oxford Movement of late 1920s as a message of hope that Rowland Hazard returned with from England to inspire Bill Wilson; at the heart of this message was the possibility of a "conversion experience," the kind of which Jung described in psychoanalytic terms as a quest for "wholeness."[20] But as fascinating as this story is historically, it leaves unasked (and thereby unanswered) two important questions.[21] First, how can we understand religion, such that its complexity can be distilled into the simplest components of a program of therapy or healing? Presumably, religion can be utilized in this way only insofar as it provides a specific way of responding to a need raised by the human predicament, so as to invoke faith in a greater power of *healing*. Second, how can the individual who is undergoing treatment become receptive to such therapy, given that the religious structure of belief is not to be superimposed by a higher authority (e.g., in the traditional guise of the clergy)?

These questions point to a larger methodological, systematic problem that our understanding of addiction must evolve, in order to keep pace with the need to develop new, complementary plans of treatment. The importance of characterizing addiction as more than an aberration, and thereby situating the addict's plight within a larger life setting, may offer simple but nevertheless important clues. Drawing upon Heidegger's pioneering insight into temporality, we find that recovery is also a developmental process. In this regard, cultivating insight into one's own personal life-situation always parallels the path of recovery. While working or executing the 12 Steps may suggest a linear progress, the development of a new self-understanding emerges in tension with the opposing tendency toward a "dissociative" misunderstanding of "who" the individual is. When seen in this light, the path of recovery presupposes the circular dynamic of temporality. For example, the adage that the addict must "choose to be sober" rings true only given the prior commitment to renew that choice (over and over again), in order to offset the destabilizing force of succumbing to the allure of one's cravings. In order to maintain sobriety, the addict must re-claim the power of choosing as such, the "choosing to choose" as the source of self-empowerment (even if buttressed by a belief in a higher power). Recovery is then the choosing again to live out each day "one day at a time." Through this temporal-spatial enactment, the self can proceed along a new *life-trajectory*, the key to any long-term healing.

---

[20] Along with Jung, William James is also cited for his account of such conversions in his *Varieties of Religious Experience*. See Kurtz 1991, p. 33. While historically James belongs to the school of American Pragmatism, in a broader sense he shared with Jung a "pragmatic" approach to the relevance of religion and its therapeutic role in healing.

[21] As in various such narratives, such "facts" are disputed, including the degree of Hazard's influence.

At some point in the future there may be a panacea to treat addiction through drug-therapy, a miracle cure or magic bullet. But until then, the mandate to live "one day at a time" rekindles the simple memory (*Mnemonsyne*) that is already intrinsic to the self-understanding of human existence. As Heidegger emphasizes, we are such finite creatures as always to have our "being to be" (*zu seinem Sein*),[22] that is, as stretching before us into the future a new challenge of taking up our existence each day. We always have our existence ahead of us as an issue, and indeed, a journey, to live out as fully as possible. In this simple observation, we find the enactment of the self-understanding that is required to nurture a possibility of healing and recovery.

1.6 Providing a Contrast to Other Approaches: The Rise of the Artifice and the Allure of Immediate Gratification

If the propensity to become addicted is becoming an "epidemic" (Drew Pinsky's claim), then there must be the potential for "hooking" individuals of diverse backgrounds. In other words, addiction would not have its global reach if the opportunity to be "hooked" were not extended cross-culturally. Despite cultural differences (and shall we add "economic" and "social" differences as well), what makes the opportunities so "equally" available? Indeed, the definition of an epidemic (i.e., like a pandemic) is that it is *de facto* "equal opportunity." There must be a further occasion, framework, and mechanism for extending these opportunities and thus making "available" whatever is potentially addictive—whether "substance" or "process." The opening up of these new channels of "distribution" requires a new configuration of the nexus between distributor and user, a "marketplace" of "supply" and "demand." If not predicated on capitalism, this nexus presupposes its own "machinations" of global success, i.e., as constructed on a new platform of selfhood and the compression of the lived-experience of time-space into micro-seconds of immediate gratification.

Here again, however, a further dimension of Heidegger's hermeneutic phenomenology provides us with a clue through his highly original and groundbreaking inquiry concerning the nature of technology.[23] Although Heidegger was among the first thinkers of the twentieth century to address the danger of technology, he did not transplant his earlier account of the existential roots on addiction on this new soil of the paradigmatic shift toward a culture of immediate gratification. By the same token, the prospect of drawing upon the full force and originality of his thinking harbors the opportunity to explore the problem of addiction from a broader horizon than can normally be achieved from any single perspective. Technology is both literally and figuratively the "engine" which drives this problem on a global scale.

[22] Heidegger 1977b/1962, p. 56; tr. 67.

[23] For a discussion of Heidegger's formulation of this question, see 1977, "The question concerning technology," in Heidegger 1977a, pp. 1–35.

Where once the opportunity for gambling was confined to the neighborhood "bookie," now the "web" of the Internet creates access for "on-line" poker and "sports-betting" of every kind. At least in the case of certain ("process") addictions, there has been an increase of frequency since the development of the Internet and, its most recent conduit, the hand-held "smart-phone." *Now the availability of the addictive source is instantaneously brought "within reach" of the addict by this technological device.*[24] The concern for "who" can become addicted becomes universally relevant, and, thereby challenges us philosophically to re-open the question of what it means to be human in the digital age. Thus, the philosophical implications for addressing the addiction become apparent, beyond its psychological impact as seen through the narrower lens of diagnosis and treatment.

Ultimately, the mercurial character of addiction highlights a broader problem of whether our capacity for compassion and self-understanding lags behind the exponential advance of technology. While "enticement" and the eventual obsession with it is one element (i.e., fetish-like character), the second element is the "showcase" thereof including the marketplace, which extends the channels of availability (i.e., a mode of "commodification"). As an important political thinker antedating Sigmund Freud (and Heidegger as well), Karl Marx coined the expression "fetish of the commodity."[25] He envisioned an "artifact," in this case a "product" of capitalism, having an additional allure over and above its explicit use-value, which ignites a *desire* for its purchase (beyond a need for it). Marx's concept of a fetish provides the historical backdrop for its introduction into the mainstream of the Western intellectual tradition. But with the transition from an "industrialized" to a "technological," digitally based economy on the one hand, and the triumph (rather than the demise) of capitalism (with its global market-place) on the other, the world has significantly changed since Marx first spoke of the fetish of commodity. In order to address this historical development, we must consider the "flip side" whereby the free market place unloosens the tethers of the fetish-like desire and now "sells" the impulse or source of addiction itself. That is, where Marx spoke of the fetish of the commodity, we will emphasize the opposite possibility of the "commodification" or the "*commercializing*" of the fetish, i.e., its global marketing through the distribution channels of technology and its *iconic-producing, imagistic-mediating power*. When the global forces of technology open up these channels, we then have all the elements to create the "perfect storm" of addiction.[26]

Given the plethora of literature on this topic, however, there have been few attempts to pause and consider the dynamic that makes addiction to be an inherently elusive phenomenon: specifically, the *double dependency* which is created by the individual's relying upon the technological trappings of modern society compound those tendencies he/she may already have to become dependent, i.e., on x, whether a substance (drugs) or an activity (gambling). Technology not only gives full reign

[24] See Schalow 2006, pp. 20–34.

[25] Karl Marx 1963, *Capital*, pp. 216–217. See Harrington 1976, p. 171.

[26] See Debord 1995, pp. 5–10. I wish to acknowledge one of my former students and philosophy major, Franklin Fehrman, for pointing out this source to me.

to the individual's tendency to seek refuge in immediate gratification, but also, by reinforcing his/her indulgences, makes it more difficult to address the addict's enslavement to them. If *denial* is the "contrary indicator" of the disposition of the addict, then the counter point is the ease by which risky behaviors become *acceptable* within our fast-paced technological world. As the philosopher of technology par excellence, however, Heidegger provides a critical counter point to our blind acceptance of the benefits of technology. Specifically, he shows that the ease with which we reap the benefits of the latest technological devices *masks the danger* of denying the accompanying risks. As an obvious case in point, the Internet becomes the conduit for the basest human concerns, as much as it opens educational and entertainment opportunities. In the narrowness of this opening, a counter tendency toward concealment prevails, the kind of subtle seductiveness that only fuels the denial by which the addict masks his/her indulgences and destructive activities.

When combined with his initial insights into the roots of addiction, Heidegger's critical outlook on technology anticipates the unique unfolding of this phenomenon in the twenty-first century. As such, his phenomenological approach stands in stark contrast to most of the contemporary, i.e., scientifically, medically, and clinically based accounts of addiction. *These approaches proceed from a paradigm of "quantification," on the one hand,*[27] *and, on the other, the "objectification" of human experience in a way that abstracts from the individual's (subjective) concern for the "meaning" of his/her existence.* In "Conversations" with the Swiss psychiatrist, Medard Boss (from 1961–1972), Heidegger differentiates between casually based, physiological models for explaining the self's maladjustment and dissociation through its deepest affects, and a philosophical approach to that phenomenon:

> The physiological dimension is an objectification of something belonging to the human being, which has resulted from a special [scientific] approach. Such an objectification cannot be reclaimed [*zurückgenommen werden*] as something characteristically human….
>
> From the fact that human bodily being [*Leibliche*] is interpreted as something chemical and as something which can be affected by chemical interventions it is concluded that the chemistry of the physiological is the ground and cause for the psychical in humans. This is a fallacious conclusion because something which is a [necessary] condition, that is, something without which the existential relationship cannot be actualized, is not the cause, not the efficient cause, and, therefore, also not the ground.[28]

By adopting a self-critical stance, our study prioritizes this dimension of meaning, as essential to a phenomenological method directed both at understanding and treating the problem of addiction.

[27] For an excellent critique of this stance of objectivism within the field of psychology, see Summers 2013, pp. 20–30.

[28] Heidegger 2001, p. 155.

1.7 The Philosophical Challenge of Addressing Addiction

Today, the study of addiction continues to be monopolized by psychology, sociology, and medicine, with philosophy having very little voice in the overall conversation concerning the origin and treatment of this vexing problem. Yet philosophy may still make an important contribution by providing an integrative vision of a multifaceted crisis, in a way otherwise lacking given the specialized focus of the behavioral sciences. Due to its wider perspective, a philosophical approach can uncover the limitations of other approaches, particularly in their efforts to compartmentalize aspects of the individual's experience of addiction, according to objective standards of diagnosis and treatment. The advantage of tailoring the study of addiction to a philosophical method that attends to the nuances of lived-experience is obvious: namely, that via phenomenology we can re-direct the concern of diagnosis and treatment to the concrete set of possibilities the individual can cultivate in the course of his/her being-in-the-world. Given this phenomenological orientation, we can redirect our study of the phenomenon of addiction within the through a unified field of human experience, which combines the spiritual, the emotional, and physical dimensions of what it means to be human—rather than prejudicially prioritizing one or the other in attempt to compartmentalize and objectify a person's addictive tendencies.

The integrative approach proposed by phenomenology avoids the many pitfalls of Cartesian dualism. Such divisions between mental and physical, the corporeal and the physical, create both metaphysical and epistemic barriers, which lead to addressing addiction symptomatically, rather then by considering the situation of the individual as a whole within a horizon of worldly and interpersonal relationships. Conversely, rather than considering the body in opposition to the spirit, phenomenology can explore the lived-body as emblematic of the openness of our being-in-the-world. In this regard, the desires and needs, wishes and dependencies, which the individual experiences in his/her encounter with addiction, are not to be considered in isolation as merely symptoms of a medical condition. Instead, these experiences re-emerge within a wider context of signifying the self's struggle to face the *pathos* of the human condition and achieve a deeper understanding of its existence. Conversely, the constellation of a higher level of meaning, and the surpassing of the self in that direction, suggests a spirituality that is interwoven into the fabric of the lived-body, rather than the opposing side of dualism set over against our physicality.

Philosophy as phenomenology enters the conversation about addiction within an integrative field, which can also distinguish the role that the individual's own choices, initiative, and, indeed, freedom can play in seeking emancipation from his/her addictive tendencies. Beginning from the first plan for treating addiction, the enigma of how a will or volition can assert itself over against its physical cravings can be viewed in a new light. For a spiritual level of self-awakening, rehabilitation, and responsibility does not arise as a separate development apart from the body, but instead presupposes the same horizon of openness within which the self can also respond to the wider spectrum of its emotions (rather than being compulsively bound by them). Because of their signifying or meaning-laden character, the self's ability to respond to its emotions defines an important therapeutic stage in its recovery, that is, understood phenomenology as the transformative moment of (the individual's) reinstatement within the wider expanse of the openness of its being-in-the-world. To be sure, there is as of yet no holy grail or alchemical formula for treating addiction, neither in the classical form of the Twelve-Step Program or more recently through advances in neuroscience. Nevertheless, there is still a heuristic sense in which the search for such a treatment benefits from guidance beforehand by a philosophically unified vision of the self, which begins from the premise of the life-trajectory of the individual in its temporal-spatial incarnality.[29]

Through the ecstatic interplay of time-space (*Zeit-Spiel-Raum*), our embodiment provides the coordinates for situating the problem of addiction within the concrete horizon of our being-in-the-world. When approached philosophically in this way, addiction is more than simply a medical aberration, but instead speaks to the larger question of what it means to be human. As Peg O'Connor observes: "Rates of addiction have been increasing along with the number of substances and behaviors a person can become addicted…. [Q]uestions about addiction are, at rock bottom, questions about the meaning of life."[30] These questions change, broaden, and deepen, however, as recent technological advances mediate human desires and thereby increase our vulnerability to becoming addicted. The current addiction crisis, then, demands that we reexamine our relation to technology, in effect self-referentially thrusting the question of what it means to be human back upon ourselves. Conversely, through our phenomenological inquiry, we will discover that this crisis (of addiction) takes shapes within the wider landscape our contemporary, within its many benefits and advantages, but also its various allures, illusions, and dangers. When seen from a phenomenological perspective, the addiction crisis speaks to the two-faced, Janus character of technology, and to the shifting landscape or topography whereupon we seek to reask the most basic philosophical questions concerning what it means "to be."

In this respect, a philosophical and phenomenological orientation breaks open the problem of addiction in a new way. As one such example, philosophy provides the encompassing vision to address how and why addiction has become a global epidemic. Not only does modern technology clear the space by which our tendency to become addicted takes root, but ultimately provides the platform to ground the human capacity to fall-prey to its fetish-like cravings. Put in Heidegger's terms, technology provides the overarching framework (*Ge-stell*) by which addiction can emerge as a ubiquitous, *universal* problem,[31] and conversely, by which we can interpret its relevance (including the possibilities) for understanding what it means to be human. By the same token, our inquiry into this vexing problem simultaneously

[29] Schalow 2006, pp. 4–7.

[30] O'Connor 2016, pp. 6–7.

[31] Heidegger 1977b, pp. 1–35. In this essay, Heidegger employs the term "enframing" to describe the encompassing matrix of technology.

challenges philosophy (and, each of us as inquirers) to address the unstable terrain or ground (*Ab-grund*), i.e., a potential fault or fissure (*Zerklüftung*) traversing Western civilization.[32] This back and forth movement between what we inquire into, and the reciprocally transforming effect on us as inquirers, further underscores the distinctive character of our study as a self-interpretive or hermeneutic venture.[33] Through this hermeneutic, phenomenological method, we will redefine addiction in a way that refers back to its emergence from the fabric of everyday life, and yet also anchors the possibility for recovery within the larger existential drama of the self's striving to achieve a measure of ownedness or authenticity. In this regard, the attempt to define addiction does not simply culminate in a succinct statement proper to a dictionary, but rather in the shift in questioning that brings to light the relevance of the phenomenon for illuminating the self's life-trajectory and its striving for transcendence.

1.8 Conclusion

While the primary thrust of this study is to re-examine the problem of addiction, its subsidiary aim is to cast light on the development of our self-understanding to meet the challenges posed by the exponential advancement of technology in our digital age. Conversely, the method of hermeneutic phenomenology that Heidegger helped to pioneer remains open-ended in its possibility to expand the horizons for *reinterpreting* his thinking, in ways that reconnect and resonate with the challenges we experience in today's world. Put loosely, hermeneutics is the art of interpretation, such that in the act of understanding, we take what we already understand and develop it anew in a completely different context. In this way, our understanding returns to a deeper level of self-understanding, whereby we appropriate previous insights in a new way, that is, *make them our own*. Interpretation, then, is the path by which we come to "own" our insights, in order that we can apply them in different contexts and thereby *expand our horizons of understanding*. In this instance, we will discover that Heidegger's thinking can still speak to the pressing issues of today, specifically, to the problem of addiction, which affects people in all walks of life. Moreover, for all its complexity, Heidegger's philosophy still harbors a simple message about the basic quandaries and questions of our human condition, as well as holds the key allowing us to address the challenges and crises of our technological and *digital* age.[34]

---

[32] See Kovacs 2015, pp. 440.

[33] Zimmerman 1986, p. 33. Zimmerman refers to Heidegger's hermeneutic phenomenology as a form of self-analysis. As a philosophical method, Heidegger pioneered hermeneutic phenomenology as an alternative to E. Husserl's transcendental phenomenology, which was cognitively based.

[34] Heidegger's philosophy revolves largely around his own innovative and groundbreaking terminology. Thus, our understanding of his thought depends on how well we can explicate his key terms. To be sure, this would be an undertaking in its own right! See Schalow, and Denker 2010,

As implied thereby, a concrete, phenomenological approach to addiction explores its "dynamics," in such a way that the nexus of self-discovery, choice, and responsibility is as important as any "psychological" or "sociological" focus on a set of behaviors and motivations, e.g., "obsessive-compulsive" tendencies, "personality-disorders," membership in a "class" or cultural backgrounds, etc. Through his hermeneutic phenomenology, Heidegger pioneered a way to address these dynamics by showing how human *temporality* opens up the horizon of possibilities within which the self chooses between alternatives. Specifically, the self can either "own up" to its existence or refrain from doing so, find guidance in the boundaries of its existence or transgress them in "destructive" acts, and, ultimately, set the *priorities* of life or undermine them in the pursuit of a single indulgence—which, in simple terms describes what addiction *is*.

In the next chapter, we will resituate the problem of addiction in terms of the larger life challenges already posed to each of us as individuals, that is, as we undertake our earthy sojourn in search of some glimmer of insight and understanding into what it means to be human.

References

Aho, Kevin. 2009. *Heidegger's neglect for the body*. Albany: SUNY Press.

Boss, Medard. 1963. *Psychoanalysis and Daseinanalysis*, trans. Ludwig Lefebre. New York: Basic Books.

Carnes, Patrick. 1997. *The betrayal bond*. Deerfield Beach: Health Communications.

Crowell, Stephen. 2013. *Normativity and phenomenology in Husserl and Heidegger*. Cambridge: Cambridge University Press.

Debord, Guy. 1995. *Society of the spectacle*, trans. Donald Nicholson-Smith. New York: Zone Books.

Erickson, Carlton K. 2007. *The science of addiction: From neurobiology to treatment*. New York: W. W. Norton.

Gadamer, Hans-Georg. 1967. On the scope and function of hermeneutical reflection, trans. G. B. Hess and Richard Palmer. In *Philosophical hermeneutics*, ed. David Linge. Berkeley: The University of California Press.

Harrington, Michael. 1976. *The twilight of capitalism*. New York: Simon and Schuster.

---

for a discussion of Heidegger's terminology. In this case, our task is made somewhat easier by the fact that the phenomenon of addiction is already familiar to most of us. As a result, certain terms will immediately jump to the forefront of our study, not simply because they are integral to addressing addiction, but, on the contrary, because its destructive character occurs against the backdrop of the larger drama of human existence. For example, Heidegger employs the term "ownedness" or "authenticity" to describe the challenge by which the self struggles to achieve its individuality over against the pressures of social conformity. On the one hand, an essential element within the dynamic of addiction is the self's tendency to succumb to these social pressures, e.g., the desire to "fit in" by "using" and "indulging" in order to be part of the "crowd"; on the other hand, only by "owning up" to oneself can a person avoid these temptations and make "constructive" rather than "destructive" choices. "Owning up to" implies the self-disclosure that is necessary for the individual "to be" responsible, e.g., to answer for (*antworten*) his/her choices. For a discussion of responsibility, see Crowell 2013, pp. 236, 302–303.

Heidegger, Martin. 1962. *Being and time*, trans. John Macquarrie and Edward Robinson. New York: Harper & Row.
———. 1977a. *The question concerning technology and other essays*, trans. William Lovitt. New York: Harper & Row.
———. 1977b. *Sein und Zeit*, GA 2, 1977. Vittorio Klostermann: Frankfurt am Main.
———. 1989. *Beiträge zur philosophie (Vom Ereignis)*, GA 65. Frankfurt am Main: Vittorio Klostermann.
———. 1999. *Contributions to philosophy (From enowning)*, trans. P. Emad and K. Maly. Bloomington: Indiana University Press.
———. 2001. *Zollikon seminars*, trans. F. Mayr and R. Askay. Evanston: Northwestern University Press.
———. 2003. *Four seminars*, trans. A. Mitchell and F. Raffoul. Bloomington: Indiana University Press.
———. 2012. *Bremen and Freiburg lectures: Insight into that which is and Basic principles of thinking*, trans. Andrew J. Mitchell. Bloomington: Indiana University Press.
Huxley, Aldous. 1968. *Brave new world*. New York: Harper & Row.
Kardaras, Nicholas. 2016. *Glow kids: How screen addiction is hijacking our kids and how to break the trance*. New York: St. Martin's.
Kemp, Ryan. 2009. The temporal dimension of addiction. *Journal of Phenomenological Psychology* 40 (1): 1–18.
Kovacs, George. 2015. *Thinking and be-ing in Heidegger's Beiträge zur Philosophie (Vom Ereignis)*. Bucharest: Zeta Books.
Kurtz, Ernst. 1991. *Not-God. A history of alcoholics anonymous*. Center City: Hazeldon.
Marx, Karl. 1963. In *The Marx-Engels reader*, ed. Robert C. Tucker. New York: W. W. Norton & Company.
Morris, David. 2001. Lived time and absolute knowing: habit and addiction from "infinite jest to the *Phenomenology of Spirit*,". *Cliometrica* 30 (4): 409–412.
Nakken, Craig. 1996. *The addictive personality*. Center City: Hazelden.
O'Conner, Peg. 2016. *Life on the rocks. Finding meaning in addiction and recovery*.
Pinsky. 2003. *Cracked: Putting broken lives together again*. New York: Harper-Collins.
Schalow, Frank. 2003. Everydayness and the question of human addiction: A phenomenological approach. *Southwest Philosophy Review* 19 (2): 17–30.
———. 2006. *The incarnality of being: The earth, animals, and the body in Heidegger's thought*. Albany: SUNY Press.
Schalow, Frank, and Alfred Denker. 2010. *Historical introduction to Heidegger's philosophy*. Lanham: Scarecrow Press.
Seeburger, Francis. 1993. *Addiction and responsibility: An inquiry into the addictive mind*. New York: Crossroads.
Stolorow, Robert D. 2013. Heidegger and post-Cartesian psychoanalysis. In *The Bloomsbury Companion to Heidegger*, ed. F. Raffoul and E. Nelson, 451–458. London: Bloomsbury.
Summers, Frank. 2013. *The psychoanalytic vision: The experiencing subject, transcendence, and the therapeutic process*. London: Routledge.
Wilshire, Bruce. 2003. *Wild hunger: The primal roots of modern addiction*. Lanham: Rowman and Littlefield.
Zimmerman, Michael E. 1986. *Eclipse of the self: The development of Heidegger's concept of authenticity*. Athens: Ohio University Press (revised).

Chapter 2
Everydayness and the "Norm" of Addictive Practices

The late 1920s ushered in a period of great social upheaval, not only in the United States, but also word-wide. Yet, that period is also one of the most intellectually fruitful, and plants the seeds across many disciplines for the study of addiction that we have today. A wave of affluence, a booming stock market, and the "roaring 20s" (at least in the United States), also created a climate of excess, exuberance, and intoxication (primarily to alcohol). While addictive tendencies were graphically depicted in the literature of the day, addiction, however frequent and troubling, was not specifically recognized *as* a problem, in the way that it is now addressed in our society. Gifted writers from America, comprising the "lost generation," would cross the Atlantic Ocean to experience an unprecedented cultural unrest and "*Angst*" that seemed to cast a long shadow over both Great Britain and Europe. Terms like "malaise," "boredom," "homesickness" would enter the cultural lexicon.[1] As we will see throughout this chapter, the development of such a cultural lexicon provides the backdrop by which an otherwise isolated, albeit destructive behavior, can be brought into question within the broader setting of the human predicament and explored in concert with traditional philosophical issues.

Not surprisingly, the sense of social unrest and uprootedness that swept across Europe in the aftermath of World War I would also provoke in adjacent philosophical quarters a new interest in addressing the most perennial of all questions, including "who am I?," "what does it mean to be human,?" and, beyond that, "what is the meaning of being?" In Germany, a youthful Martin Heidegger pioneers a new methodological twist on the formulation of these basic questions, that is, by developing his approach of hermeneutic phenomenology to interpose a concrete orientation of self-inquisitiveness into these traditional concerns. Yet, interwoven into that approach is a *concrete, groundbreaking terminology that would change how*

[1] Heidegger 1983/1995, pp. 6–10; tr. 5–7. Homesickness as a grounding or fundamental attunement (*Grundstimmung*) becomes crucial for Heidegger in this lecture course from 1929/30, and for decades subsequently.

F. Schalow, *Toward a Phenomenology of Addiction: Embodiment, Technology, Transcendence*, Contributions To Phenomenology 93,
DOI 10.1007/978-3-319-66942-7_2

subsequent generations would both think and speak of the human predicament. In a way that might otherwise get lost in his voluminous text, *Being and Time*, Heidegger provides one of the first philosophical descriptions of the self's potential to "spiral downward" into the trap of addiction.

The confluence of cultural and intellectual horizons also proves to be fortuitous in the attempt both to plant the seeds for a future attempt to understanding addiction, and in the current challenge of recognizing it *as a problem* in order to spark a search for a possible plan of treatment. Historical speaking, in this specific instance the pragmatic thrust of seeking such a plan would evolves, as it were, in inverse relation to the development of an intellectual tradition from understanding and addressing that whether in philosophical, religious, or psychological terms. In this respect, perhaps the greatest irony in the development of the first program for treating addiction is that its founder, Bill Wilson, had no professional expertise: he was neither a psychiatrist nor an M.D. (like his father-in-law); nor had he any training as either a theologian or a philosopher. And yet, his legacy is that, through the movement and treatment he originated, Bill Wilson would probably have greater influence than perhaps but a few women or men of letters and advanced degrees in the twentieth century. A man of unremarkable qualifications, an average stockbroker (and perhaps not a very prophetic one on the cusp of the great stock market crash of 1929), Bill Wilson rose to remarkable heights of achievement. Yet, this simple *fact* will not only be important in helping us to situate the problem of addiction. Emphasizing this fact will also provide a key for earmarking Martin Heidegger's innovation in the development of his phenomenological method, i.e., as *incorporating a personal, self-questioning response to the crisis pervading human existence*. For prior to any program of treatment that an expert may propose, there is the thread of *everyday* experiences joining addict and non-addict alike, and which must be factored in as the *backdrop* for developing a genuine understanding of the *problem* of addiction. Within the more technical confines of his philosophy, Heidegger coined the term "everydayness" to describe the point of departure for examining human existence and outlining the structures definitive of it.

Analogously, in developing the question of addiction, we need to identify a new point of departure from that which has ordinarily been assumed in most studies of this problem, particularly given the precedents set in such fields as psychology and sociology. Most studies accept the obvious premise that addiction is an "aberration," implying certain kinds of deficiencies from psychological defects of low self-esteem to physiological imbalances in a person's brain chemistry. Not only do most so-called "scientific" studies assume a Cartesian split of mind and body,[2] they also ignore the common existential questions that challenge addict and non-addict alike. These are the basic questions that express the existential predicament which each of share as being-in-the-world. Rather than fragmenting the problem of addiction, it might be more fruitful to re-examine its emergence within the larger continuum of life, in which the individual's endowment with the capacity to care shows that the

[2] For an important attempt to circumvent this mind-body dualism, and phenomenology's contribution to this end, see Aho, James and Kevin Aho 2008, pp. 26–29.

impetus for overcoming the problem lies first in the self-awareness of identifying it *as* a problem. The dynamism and growth of the individual's awareness, as a permutation of what Heidegger calls "owned" or "authentic" existence, cannot be underestimated as we proceed to address the problem of addiction. For only by establishing what it means for the self to assume ownership of its existence, can we, by contrast, address the opposite possibility whereby the individual succumbs to his/her addictive impulses and desires. By making the individual's concrete situation the point of departure for our inquiry, equally in his/her possibility of growth as in the opposite possibility of degeneration, we transpose the focus of the study of addiction in the direction of examining, in Heidegger's terms, the phenomenon as such (rather than its "symptoms").

Yet, in order for our phenomenological study to bear fruit, we must also situate it within the larger historical context from which the discussion of the problem of addiction first entered into the mainstream culture. Though human beings have been vulnerable to different kinds of addiction since the dawn of civilization, its emergence as a problem meriting study and discussion coincides with the figure who first began to document specific forms of addiction, e.g., alcoholism, *as* a problem, namely, Bill Wilson. Indeed, the full context for questioning the roots of addiction does not emerge until we simultaneously incorporate a historical perspective on this problem By the same token, it is not merely as historians that we do this, but rather, in concert with Heidegger's own hermeneutic method, we approach Bill Wilson as a concrete figure in search of self-understanding, as well as the pioneer of the first program for treating addiction. Accordingly, in following the guidance of Heidegger's hermeneutics, which transposes the focus for studying addiction, we must ask first "*who* is Bill Wilson?," rather than "what he did?" In asking this question, we further shift the inquiry back to the level of the concrete situation of the individual, or what Heidegger calls the "factical" circumstances of life.

In this chapter, I will point to Bill Wilson, the individual, as a foremost example of how the existential predicament of the addict can illustrate, or in Heidegger's terms, "formally indicate" the pre-philosophical, pre-scientific background that we all share—addict and non-addict alike—out of which any genuine understanding of the problem of addiction can first arise. Following the basic precepts of Heidegger's hermeneutic phenomenology, I will, secondly, identify "everydayness" as the point of departure for our study of the *phenomenon* of addiction. Given this phenomenological orientation, I will, thirdly, outline the existential preconditions for the self's developing an extreme "dependency," either on substances, e.g., drugs, or other activities, e.g., gambling. In this regard, we will discover that everydayness points to *1) the origin of the self's tendency to become addicted, 2) the life-situation or existential context for understanding addiction, and 3) the dynamics of falling, including the self's interaction with its environment, which creates an atmosphere as well as an opportunity to become addicted.*

2.1 The Everyday Predicament of the Individual

As mentioned in our Introduction, the German word *Zeitgeist* or spirit of the times describes the historical-cultural milieu wherein Bill Wilson first confronted his problem of addiction. In philosophical terms, the spirit of the times corresponds to what Heidegger describes as the basic fact of human existence as always emerging within a specific temporal horizon and historical period, and thereby as already "situated" in the world. A perhaps more graphic and poetic description that Heidegger also employs may also be illuminating, specifically, in defining the self as embarking upon a "temporal sojourn," the journey of its "being-in-the-world."[3] In an analogous way, Bill Wilson's legacy lies in his unique journey of self-discovery, in and through the vicissitudes of addiction and the gateway of insight (into his problem) opened up thereby. Yet there is another level we must first consider, as a pathway to explicating Heidegger's basic philosophical concepts, and that is how the uniqueness of Bill Wilson's historical milieu can yield clues to "situate" our discussion of addiction. For Bill Wilson's historical time and place is the turning point in the transition to a world-economy that is based on buying and selling, trading and consuming.

For the one "accident" of history we cannot discount, as it were, is Bill Wilson's unique occupation for his time and place, that is, as a stockbroker on Wall Street in New York City in the 1920s before the market crash. For it is not only the case that he lived in a culture of "excess" in the decade of the "roaring 20s," but that "live for the day" attitude emerges against the historical backdrop of the death and destruction of the "war to end all wars." With this example we begin to see that the so-called "environmental" influences that shape the individual's tendency to become addicted also include a nexus of cultural and historical factors that create the atmosphere in which this kind of problem can flourish. Indeed, in looking at this larger picture, we begin to see that whatever social and personal factors we can isolate, their degree of influence becomes magnified when integrated into the "macro" forces of this culture of excess. For what this culture of excess breeds, and for which the world-economy of consumerism becomes the stage, is a license for, and exhortation of, a life-style of "immediate gratification."

Though there is an obvious link between this life-style and the proclivity toward addiction, we need to step back and outline some other key motifs before developing this correlation. For what is at issue is not only an excessive life-style, but what it also *signifies, by way of contrast*, concerning other keys to living a full, enriched life which may easily go unnoticed. From Heidegger's perspective, we need to emphasize that the possibilities one is granted are the foremost measure of the abundance of life and the impetus for acknowledging one's gratitude for simply existing at all. Heidegger coined the term "Da-sein" to describe the simple awareness of the fact of existing, of "being-there," through which each of us as individuals are disclosed through one's unique possibilities. Though we need not enter into a technical

[3] Heidegger 2001, p. 87.

discussion here, it is important to mention that, according to Heidegger, human existence can be developed in either of two ways, either as "owned" and individualized or "authentic," or as "unowned" and divested of individuality or "inauthentic." In the former case, owned existence involves developing one's unique possibilities, apart from any pressure of conformity; while in the latter case, unowned existence entails adapting one's possibilities in order to "fit in" and adhere to the "they-self" or the crowd. Put simply, in "unowned" existence the individual compromises his/her individuality in order to fulfill the expectations of others.

As we will discover, the development of this contrast between the "owned" and "unowned" ways of existing will not only go a long way to uncovering the roots of addiction, but will also suggest keys to the process of becoming aware of and accepting the problem, as well as providing the impetus for a commitment to change. In this regard, we must re-emphasize that the owned and the unowned, authenticity and inauthenticity, pertain to how the individual stands within the world. Indeed, for Heidegger human identity is not merely a facet of one's personality or of character traits, but instead also includes the compass of relationships which include interaction with others, as well the use of, and dependency upon, things within the immediate environment. The tendency to describe addiction as a "weakness" of character or a personality "flaw," then, abstracts from the greater complexity of the relationships that are the hallmark of "unowned" existence. Specifically, the divestiture of individuality which occurs in unowned existence also translates into a model of treating others, that is, in a parallel way to the instrumentality by which we use things, and, conversely, the self's stance to the world is constricted to its dependence on the use-values of things and the satisfaction of its desires accordingly. As a permutation of inauthenticity or unownedness, addiction distinguishes an extreme form of dependency, the possibility of capitulating to whatever satisfies one's craving for immediate gratification.

In this regard, the sense of the "immediate," by way of contrast or as a "contrary indicator," suggests that the experience of time, "lived-time" or what Heidegger calls "temporality," will ultimately define the demarcation of human existence through the dual paths of the "owned" and "unowned" self. Heidegger's account of temporality, which holds the key to his entire philosophical task, is as intricate as it is provocative. But if we were to single out one motif that is as central as any, it would certainly be that human temporality is circumscribed by its finitude. Indeed, as temporally defined, the self always encounters its possibilities as limited, not only by the circumstances of its situation, or what Heidegger calls "facticity," but also by the inevitability of death. But for him, death is not merely an event that lies "at" the end of life, but as a possibility in its own right, is that (end as) finality *toward* which each of us is already directed and which, as such, puts into perspective all of my other possibilities. In this regard, the negativity of death is linked to the eminently positive dimension of highlight the uniqueness of my possibilities. Why? Because from Heidegger's standpoint the acceptance of the limitation of death 1) induces a sense of awareness about the importance and urgency of the "time" allotted to me and 2) summons me to identify, establish, and act upon specific priorities that give direction to my life, providing signposts to mark the development of my individuality.

Turning back again, to the opposite, we discover that in its finality death also indicates, by way of contrast, the beginning of the tendency of the individual to seek escape in addiction. For in its initial phase, or before it has become irrevocably an avenue of such escape, addiction offers the allure of removing the individual from the constraints of all limits. And these limits, from the standpoint of the "fugitive" path of escape, pertain to both time and place: as uncontrollable circumstances that hold one in a state of distress or anguish (from the past), and as the promise of a continual satisfaction of one's desires (in the future). Only an extremely truncated form of the present remains open as an avenue to act on one's desires, if only indirectly as a "motive" for escape. Hence, because addiction begins from a tendency all human beings have to negate "deny" their limits, the focus of concern shifts exclusively to what offers satisfaction in the proximity of the present. Hence, the impetus toward immediate gratification, which plays a prominent role in any addiction, is more than just a cliche, provided that we construe this specific factor within the overall dynamic of temporality as Heidegger describes it.

It is only because the possibility of setting limits is central to human existence that problems can arise due to an opposite pursuit of self-indulgence and excess. In the culture of excess, which emerged in Bill Wilson's era and is even more pronounced today, we tend to review limits or boundaries as restrictions. But when we recognize, on the other hand, that the first step to self-awareness lies in acknowledging one's limitations, a completely different perspective emerges. Rather than simply standing for obstacles, one's limits can also implicate "balance" and "equilibrium." While such equilibrium may at first suggest a "state of mind," it also includes an active component of the power of balancing priorities. To be sure, there is a temptation to oversimplify the problem of addiction, which mistakes the negative effects for its cause, specifically, the examples of "lack of control." But the "lack" is not the root or "cause" of the self-indulgence. Rather, we must consider where the source of self-empowerment first arises, which matters of the will and "control," or the lack thereof, ultimately presuppose. In this way, we continue to transpose the focus of our study away from merely observing the negative behavior, and extrapolating to its so-called "proximate" causes. Instead, we attempt to reexamine the larger context and "life-setting" in which both the possibility of the problem arises, as well the possibility of coming to terms with, and becoming aware of, the problem *as* a problem.

With this observation, we arrive at the key to applying Heidegger's hermeneutical method. The method of study arises from, and points back to, the self-understanding which sustains an awareness of the problem, the "as structure" for its determination. Indeed, addiction is a specific example of a problem whose recognition *as such*– for the individual as well as society–is particularly elusive. Conversely, the need to do so illustrates the hallmark of hermeneutics, that is, its mission to elevate what is otherwise elusive, whether from the perspective of everyday life or philosophically, to the level of self-understanding. In the development of his hermeneutics, Heidegger sought to formulate from the (act of) individual's self-understanding, the basic concepts for philosophically addressing the structure and "meaning" of human existence, and, ultimately, of being. In employing Heidegger's hermeneutic method to study the phenomenon of addiction, we embark upon a similar path. For by emphasizing the importance of establishing boundaries and priorities, as a counterpoint to the culture of excess and instant gratification, we point to the basic concepts that are germane to our study of addiction.

As we will see, the discourse that duly unfolds occurs in tension with the "objectifying" language of the sciences, which ordinarily makes the study of addiction their special domain. Foremost to implementing this hermeneutic strategy, we need to address the role that "everydayness" plays in allowing us to describe the confluence of factors that make an individual vulnerable to becoming addicted.

2.2 Everydayness, Conformity, and Peer Pressure

As we proceed with our study, we will witness several changes of viewpoints in regard to what is conventionally believed about the nature of addiction. And one of the key strategies of Heidegger's hermeneutics is to develop a completely new insight by showing that it is hidden in what we ordinarily take as obvious or held by convention. One such example arises when we move from the apparently obvious inference that, as a negative behavior, addiction is something "abnormal" when, as we shall discover, the inverse is more precisely the case: that from the perspective of the culture in which we live, addiction is rather an example of "normalcy" viewed in a larger, philosophical sense. And, of course, this distinctive sense corresponds to what Heidegger describes as the routine of our day-to-day lives, a kind of paradigm of regularity that he calls "everydayness." Put in the simplest terms, everydayness defines the context of our life-regimen in which the "averageness" of understanding, the predilection to "fit in," and the various modes of socialization prevail. In this rather undistinguished realm of human existence, the tacit acceptance of the "status quo" offers the illusion of both comfort and security, as well yields the atmosphere in which individuality may be compromised in favor of the desire to "conform" and adhere to social norms or conventions.

Not only will everydayness situate the initial phrase of our analysis, it will also point to the "persona" of addiction, that is "any man" (or woman), of whom Bill Wilson, the ordinary stockbroker is the foremost example, and also potentially everyone. The "everyone," however, does not mean that each person is equally predisposed to become addicted, but that rather than addiction always emerges within a social setting. While the tendency is (and perhaps this is one of the distortions of the "everyday" mentality) to view addiction as an individually rooted problem, the facts speak quite to the contrary. For as "being-in-the-world," as Heidegger emphasizes, each of us is already being-with others. As a result, addiction emerges within a certain cultural-social climate, which preys on the individual's insecurities on the one hand, and, on the other, allows family, friends, and colleagues to fuel the problem by virtue of their own preoccupations and even indifference. If we can point to a distinctive mentality of everydayness, and its expression in our fasted-paced, technological world, it is the simple way in which the individual becomes so "wrapped up" and "preoccupied" in his/her pursuits as to avoid seeing, let alone addressing, the emergence of a more serious problem.

The distorted perception of Bill Wilson's time was that the addict lived an isolated existence, the "skid row" bum, the exception and not the rule. The simple appreciation of the fact that the individual occupies a social network in "being-with others" suggests quite the opposite: that everydayness allows for the confluence of conditions whereby we occupy a cultural milieu along with those struggling with the problem of addiction. As a counter example, while there is obviously a criminal element within (American) society, we do no actually interact directly with them. But the tendency to become addicted, in the rich and the poor, the young and old, even the religious and the non-religious is a pervasive element to it that almost "everyone" comes in contact with the problem sooner or later, e.g., the professor whose job it is to "teach" the student who comes to class "stoned." Perhaps there are sociological reasons we can extrapolate for this pervasive problem of addiction, for example, the breakdown of the family, etc. Indeed, such approaches may prove illuminating, provided that they incorporate the wider historical perspective of our *technologically based culture of instant gratification*. But regardless, we cannot fully understand the interaction of the forces without also addressing the elements from which they emerge through the "mind-set" of everydayness.

The elements that are unified within the structure of everydayness, which most significantly impact the individual's tendency to become addicted, are "conformity" and "complicity." These two elements reside within the unity of what Heidegger calls the "they" (*das Man*) or the "crowd." In terms of the problem of addiction, "conformity" translates into "peer pressure," while "complicity" translates into "co-dependency." The difficulty with these latter terms is that they not only harbor certain connotations within the popular lexicon, but they also have a precedent of usage within the behavioral sciences. As a result, we need to outline the experiential dimensions of these possibilities as they arise from everydayness. Indeed, the "they-self" personifies that attitude which is ubiquitous throughout society and thereby opens up the "trap" of addiction into which any individual may fall, the common place desire to conform, to "fit in," to be accepted, to belong, to be "cool," even at the expense of compromising the self's individuality.

Addiction would not be such an easy "trap" in which to fall if its activities did not already have a tacit consent and acceptance accompanying them. Even if there may be laws in place to prohibit these activities, whether "prohibition" laws a century ago or drug laws today, there are still pockets within society that lend their approval by welcoming them within a "circle" of "fitting in." Because the "they" defines the self as "undifferentiated," there is a necessary indeterminacy to it whereby the sphere of approval can correspond to the "mainstream," as smoking cigarettes once was, or occupy the "fringes," as with the use of "ecstasy" among participants at the "raves" in the 1980s. But what the reach of this sphere of influence is, the price of conformity is always compromising one's individuality. The individual thereby discovers his/her identity, not through a set of unique possibilities, but instead through those possibilities that belong to and represent the aims of the group. In extreme cases, the individual may even define him or herself through the specific behaviors, which a group exemplifies as a group, i.e., "partying," "getting high," "risk-taking,"

Yet, as far as addiction goes, the "they" poses the greatest danger in serving as a mask or persona behind which the individual may hide in his/her indulgent search for pleasure and escape. For in identifying with the persona of the group, the individual becomes removed from his/her initiative both to think and act separately and to own up to what is occurring. In this way, the individual can easily skirt the boundaries which give balance and direction to his/her existence, and thereby not only succumb to over indulgence; but, in the process, capitulating to the group has a disproportional effect of also removing any *measure* of "good sense," to the point of divesting the individual of his/her ability to "own up" to, and take responsibility for, his/her deeds. In terms of the language of the vernacular, the divestiture of individuality involves a complete abandonment to risk, a loss of all inhibitions. As a result, the individual may not only pursue self-destructive actions, but put others at risk as well, e.g., as in driving under the influence of alcohol. In this respect, "peer pressure" is not only a key factor in "initiating" an individual into a potentially addictive activity, but also in "pushing the envelope" of its risk to new extremes. Accordingly, the taunts of "daring" serve as excellent examples, which pervades the culture of many college fraternities.

Yet, the impetus to conform constitutes only one side of the problem, insofar as the "they" also shapes the contribution of the "non-addict" in allowing a climate of addiction to flourish. For the complicity of the non-addict goes hand in hand with the conformity that gives impetus to, and license of approval for, the potentially addictive activities. That complicity lies in overlooking the degree of the other's involvement in a specific activity, as a pretext for doing nothing. The pretext may be as innocent as simple preoccupation with one's own concerns, or it may be more explicit through an active role of making excuses, such as, "everyone drinks on occasion," or over the holidays, or when one is down. What is most revealing in such locutions is the anonymity of the referent "everyone." For in such innocuous remarks the voice of the "they-self" resounds, as a way of blurring the lines where responsibility lies, either on the side of the individual who is in complicity or on the side of the addict. Indeed, the everyday level of this complicity explains the peculiar fact of how, at least in the initial stages of their addiction, many addicts continue to function as "pillars of society."

The hallmark of this complicity lies in the willingness and propensity to make excuse for the destructive behavior of the addict. Of course, the term "enabling" is the buzz word that has entered into the popular lexicon, as much as it is use most psychological studies. Yet that is one term which I would rather avoid using here, because it suggests a mode of potentiality when the phenomenon it supposedly describes involves just the opposite, or a kind of "privation" and diminishment of one's ability and potential. We might be better served to suggest that the complicity gives absolution to the addictive behavior, and thereby encourages its continuance, primarily as an "omission," rather than as a "commission." The so-called "enabling" depends upon another's "refraining," which characterizes the predominant *modus operandi* of the "they" of preserving the status quo. Though we will later take a closer look at this phenomenon, the addict and the ("significant") other may co-exist because of the latter's desire to preserve something beneficial in the relationship or even out of fear of loneliness. Yet this complicity reveals not only that there is a social component to any addiction, but, even more importantly, the pervasiveness of the influence of the "they" of sustaining the inertia of the "status quo," of resistance, reluctance, and indifference. Without the pervasive inertia of everydayness, there could be no climate for addictions to flourish.

By appealing to everydayness, we are able to examine the entire context of involvements and relationship that are interwoven into the problem of addiction. We are thereby able to appreciate the unique dynamics that pervade the problem, specifically, that despite their differences, addict and non-addict may still be joined by a common thread. For what begins as minor compromises and concession, which play into human existence with everyday regularity, may spiral outwards to shape one's way of interacting within the world. We allude to the horizon of the world in order to emphasize that this tendency to mask and masquerade, to mislead and even deceive, is not merely an internal coping mechanism, but overshadows the self's entire being-in-the-world. Heidegger describes our vulnerability to this duplicity "falling," which defines the inertia of our preoccupation and absorption in our everyday concerns. While various indulgences can come and go, they could never develop into an ultimate dependency without one crucial element of the human condition: deceit, duplicity, dissimulation or however we might characterize this phenomenon. While there may be many recreational pursuits that appear addictive, what makes them become so is that they suggest a "habituation" which gradually rises to the level of a "dependency." Yet such complete dependency could not be sustained, let alone initiated, without the tendency toward deception and dissimulation which lies at the root of falling and allows for its "intensification."

In simplest terms, addiction is a possibility that plays out within the shadows of deceit and deception. With the widening and deepening of these shadows, addiction becomes a downward spiral ("*Absturz*") into a bottomless pit. As Parvis Emad and Kenneth Maly emphasize, the use of the prefix "ab" has special meaning for Heidegger, by emphasizing the refusal and staying away of any support or ground.[4] And this characterization is particularly *apropo* to describe the scenario of addiction, which, as the layers of deception begin to accumulate, implies an almost impenetrable obfuscation from which it is difficult for the addict to extricate him/herself. For an addiction begins to form its roots when the individual initially disguises the motivation for doing X, e.g., "I need a little pick-me-up in the morning," or "a drink to help me unwind in the evening." And from this initial dissimulation another level forms, e.g., through apologies and excuses, and then another level, e.g., by concealing one's behavior from others, and still another, e.g., by denying to oneself and others the extent of one's dependency, and so on even to the level of

[4] See Parvis Emad and Kenneth Maly 1999, "Translators' Foreword" to *Contributions to Philosophy (From Enowning)*, p. xxxi.

criminality, e.g., of stealing to get money in order to buy drugs and cover gambling debts. As one level of dissimulation piles on top of another, we see how an addiction can become both destructive and difficult to overcome.

Indeed, the hallmark of an addiction, in contrast to other behaviors that may also exhibit a similar tendency toward self-indulgence if not "deviancy," is that dependency seems to acquire its own momentum. This is self-sustaining, downward momentum is the lived-experience of what Heidegger calls "intensified falling." The danger posed by such falling lies not only with the allure of the addiction, but also with the fact that it gains its own impetus *by simultaneously subverting the individual's own ability both to recognize and counteract the dependency*. This tendency toward self-subversion fuels the falling of addiction, in such a way that it taps the energy of one's self-concern and consumes one's "well-being" in service of sustaining the addiction at any cost. At this crucial juncture, the sense of "balance" and "discretion" to which Heidegger attributes to the ultimate constitution of human existence as "care" has been severely compromised. But even more catastrophically, in the self-subversion, the whole dimension becomes "inverted," such that the drive to perpetuate the addiction clashes with even the most primitive "instincts" of survival. The layers of deceit extend so deeply as to show, ironically, that the *true indicator of an individual's addiction lies in the vehemence of his/her denial.*

The manner in which this denial and dissimulation can become so entrenched explains why addiction can prove to be such an insidious problem, and induce "depravity" on a social as well as an individual scale. The question has been asked whether left to their own resources animals show any inclination to become "addicted." Yet it might be more appropriate to ask whether the converse, namely, what unique capability do human beings exhibit, in forsaking truth for its opposite, which underlies the possibility of any addiction? And the answer, as we have already suggested, is that human beings are equally capable of concealment and covering up as they are of disclosedness and self-discovery. As Heidegger remarks in a provocative passage in his 1929/30 lectures on the nature of animals: "No animal can sink as low as man."[5]

We can illustrate this depravity by conjuring up the images of an addict who has sunken to the level of debauchery. But the lies and deceit are perhaps a greater source of offensiveness than even the obnoxious behavior and habits of many addicts. For the all too familiar manner in which deception casts its web insures that other people, along with the addict, may become ensnared and suffer as well. Within the context of everydayness, the simple fact that addiction is common place, and thereby is a problem that "strikes at home," indicates why it is extremely difficult to ignore. And, as Heidegger suggests, the manner in which addiction speaks to the vulnerability of our human condition, and hence is a problem that is as "close" as the glasses on our faces, also shows why it is profoundly difficult to bring into focus *as* a problem. For its own part, hermeneutics holds great promise, because it provides a strategy to peal back the "closest" levels of self-understanding, in order to

---

[5] Heidegger 1983/1995, p. 286; tr. 194.

arrive at the dynamics of human existence, which simultaneously harbor the roots of addiction and provide the keys to cultivate a genuine awareness of the problem.

As we have seen, the phenomenological approach to addiction opens up an avenue for studying the problem, which diverges sharply from other approaches. Put simply, hermeneutics addresses the problem of addiction at the root of its factuality, as a dimension that is already shaped by the conventions, habits, and practices, which in their familiarity to us all form the texture of "everydayness." Other methods arrive on the scene after the fact, and thereby, from an abstract standpoint, construe addiction as an "aberration" of a normal state of being. Our discussion, however, has revealed a completely different dynamic to the problem, in which the deceptiveness of a fugitive existence (i.e., "falling") sets the "trap" of addiction, into which the individual plunges through layers and layers of deceit and dissimulation.

Having established everydayness as the overarching context from which to study the problem of addiction, let us conclude this chapter by examining how "falling" can create the specific conditions of "dependency." In the process, we will also address the link between dependency and habituation, including the existential basis of the latter.

2.3 Human Existence and the Existential Preconditions of "Dependency"

In his magnum opus, *Being and Time*, Heidegger outlines the structures of human existence as "care" (*Sorge*). According to him, human existence, as situated in the world, allocates space, both as a place where it dwells and as a measure of nearness and proximity by which to direct its everyday use of things or items of equipment, e.g., a hammer. Likewise, temporality allows the self, by projecting the future, to emerge within a horizon of possibilities, and, by rediscovering the past, disclose its uniqueness through its actions in the present. As conjoined with each other, temporality and spatiality animate and shape all human concern, even its distortion and deprivation in various addictive pursuits. If temporality and spatiality ultimately constitute human existence in this way, then they must also outline the conditions whereby the momentum of falling takes over and shifts the self's preoccupation to whatever in the immediate environment becomes an exclusive focus of concern, whether a "substance" (alcohol) or a pursuit (gambling).

In accord with Heidegger's hermeneutic methodology, I appeal to temporality and spatiality as preconditions of addiction, in order to describe the dynamics inherent in this mode of intensified falling, rather than to identify any specific "causes." In an analogous way in which Kant first distinguished time and space as the hallmark of human finitude, these preconditions interplay in a concrete manner to restrict the horizon of the addict's self-awareness and his/her involvement with the world. Insofar as "falling," along with facticity (the pregiven set of circumstances) and ex-istence (the standing forth in possibilities), defines the essential structures of care, then the self's propensity to "fall" narrows the focus of its awareness and concerns. As an everyday occurrence, which implies a comportment toward oneself (if only indirectly) and to the things we encounter in the world, falling unfolds on two fronts. First, falling is a distinctive expression of care, even in the indifference of not caring, directing attention to whatever emerges in one's environment as interesting, alluring, and even as "tempting." The diversion of care in this way takes the form of "wishing," "wanting," "desiring,"[6] and, in the most extreme cases pertaining to addiction, of "craving" something which offers the possibility of immense gratification. Secondly, in diverting attention to things, falling takes its cue from the "they-self," such that the pursuit of gratification goes hand in hand with the urging, appreciation, and approval of other people who validate these diversions. As a result, the weakening and compromising of an individual's sense of self-identity gives further impetus to acting impulsively upon one's desires. We need only to recall the "Animal House" environment of the college fraternity as a good illustration. In the nexus of the double movement of falling, where the peer pressure of the "they" converges on the craving for what environmentally poses as a source of gratification, the possibility of a fixation arises, in short, an addiction. As a simple example of this dual movement of falling, we can point to an individual who seeks to win the approval of others, and thereby bolster low esteem with the intoxication of the "high," by compulsively shopping for "flashy" clothes.

Within the context of everydayness, falling already implies spatiality in terms of the "proximity" of the things within the environment we regularly encounter. The "environment" pertains simply to the "surrounding" realm of concern, within which whatever we encounter becomes significant in relation something else, e.g., as tool or item of equipment, so as to form a chain of involvements. Insofar as the encompassing horizon of these involvements defines the world, or our situatedness within it entails that we exist already "alongside" things and equipment, we also co-exist with others. Because of their proximity, the things "alongside" us are characterized in a distinctive way, that is, as "ready-to-hand" insofar as they serve our instrumental designs through their "availability." As Heidegger states in his 1923 lecture course on the hermeneutics of faciticty: "This being-ready-to-hand, being-at-our disposal, constitutes its *availability in advance.*"[7] If we consider the constitution of the environment in terms of the everyday preoccupation with getting things done or work, the so-called work-a-day-world designates items of equipment as available, as "within-reach" or ready-to-hand. Spatiality emerges as the dimension that prestructures our access to the environment, specifically, to the work-world, in terms of the demand of convenience. Space is not initially geometrical or equivalent with the extension of physical objects, as Immanuel Kant saw fit to explicate in the Transcendental Aesthetic of the *Critique of Pure Reason*. On the contrary, environmentally lived-space gauges the relative closeness and farness of whatever we

---

[6] Heidegger 1977/1962, p. 258; tr. 238.

[7] Heidegger 1988/1998, p. 97; tr. 75.

encounter and then negotiates the distance to achieve proximity, availability or the "within-reach" of the ready-to-hand.

The term "availability," as it becomes central to our study of addiction, has a twofold meaning, first as indicating the measure of an individual's access to something and, secondly, what might be understand as a social outlet for using or doing something. As is somewhat obvious, the relative degree of "availability" to whatever can satisfy and yet fuel one's cravings, seems to be an important factor in the development of any addiction, both on an individual and a social scale. Indeed, making drugs available is certainly a large factor in why so many people become addicted to them, and hence the so-called "war on drugs" aims at preventing their production and distribution throughout the United States. In phenomenological terms, however, availability does not simply refer to making a given activity/substance easy to get. Instead, availability refers to the impetus to place "within-reach," such that the addict is always seeking proximity with whatever can satisfy the "addictive thirst." The impetus to bring within "within-reach," however, implies a unique dynamic of "spatializing" or "bringing close," which Heidegger calls "deseverence." To "desever" is to reallocate the space, in such a way as to take something that is initially far or "outside" of reach, and, as it were, bring it "within-reach" in terms of magnifying something's *availability*.[8] To desever (*entfernen*), then, is to relocate something we encounter environmentally and bring it within the scope of availability, thereby rendering it proximal or easily "within-reach."

From the standpoint of the existential preconditions of addiction, such desevering is essential to the *modus operandi* of the addict. Indeed, if we connect this desevering to the facticity of the individual as spatially located, as incarnated or embodied, then this reallocation of space, qua addiction, translates into making the substance or activity almost a physical extension of the individual. The best example of this scenario would be cigarette smoking, in which the wrapped tobacco becomes physically intertwined with the person's lips and hands. We all know cases in which the individual rarely appears without being seen with a cigarette either being smoked, lit, or handled, for that matter. And the key to overcoming such an addiction is not just conquering the physiological dependence on nicotine; the problem also hinges on altering the manual habit, which the smoker has acquired, e.g., in lighting and handling the cigarette. The concreteness of the hermeneutics of facticity enables us to see that the physicality of addiction includes as much the bodily extension of the within-reach (i.e., the drink or cigarette as a second appendage), as the biochemical processes of either the substance (i.e., the metabolism of alcohol or the adrenalin rush triggered either by computerized day-trading or video-poker).

Is not addiction, then, just a subset of what psychologists describe as obsessive-compulsive disorder? While this claim may be accurate, when offered an explanation it seems to do no better than to beg the question. For how does such a disorder originate in the first place? At best, this psychological appeal once again falls back upon the presumption that addiction is a kind of aberration. When we characterize addiction as a fixation, however, that implies some point of fixity, which in turn

[8] Heidegger 1977/1962, pp. 140–141; tr. 139.

suggests a specific reallocation of space within which the addict experiences the world. Insofar as the desevering in which the addict engages brings the addictive source within-reach, brings it into the foreground as the source of his/her craving, other items of interest slip unnoticeably into the background of attention. While working at a country club, I once saw an inebriated golfer fall after having tipped a golf cart he was driving. Much to my amazement, the golfer's first concern was not his skinned arm, injured during the spill, but locating his can of beer! The fixity of addiction is such that it *displaces* everything else worthy of attention or concern into the background, thereby creating a vacuum in which only the within-reach can emerge as having significance. Though Heidegger was not specifically addressing the "clinical" problem of addiction as we know it today, he provides a remarkable description of the existential basis of that problem as an extreme form of "hankering," wishing, or desiring:

> Dasein's hankering as it falls makes manifest its *addiction* to become 'lived' by whatever world it is in. This addiction shows the character of being out for something [*Ausseins auf....*]....What one is addicted 'towards' [*Das 'in-zu' des Hanges*] is to let oneself be drawn by the sort of thing for which the addiction hankers. If Dasein, as it were sinks into an addiction then there is not merely an addiction present-at-hand, but the entire structure of care has been modified. Dasein has become blind, and puts all possibilities into the service of the addiction.[9]

In simplest terms, this "blindness" narrows one's focus or perspective, which in the form of intensified or "aggravated" falling translates into the fixity we see in addictive behavior. According to Heidegger, human existence has the essential tendency, corresponding to falling, toward concealment, dissimulation, and deception. But the self also has the opposite capability of standing outside and beyond itself toward possibilities, of entering into the openness, as it were. As Michael Zimmerman points out, falling is a condition that "contracts" this ontological openness.[10] With the contraction of this openness, an ontical "craving" ensues, which displaces interest in everything else in favor of what offers the promise of immediate gratification, the within-reach. In alluding to an intensified falling, Heidegger points to the accumulation of layers of concealment in which the self becomes entangled, thereby placing the individual, as it were, on a slippery slope, the vertigo of a "*downward spiral*." All of these descriptions are helpful, but what examples of behavior do they translate into which define the fixity of addiction? The answer lies in the way that the indifference and indecisiveness of the "they" lurks in the background of the self's succumbing to the temptation of the facile source of its fulfillment, of transposing its identity into the narrow locus of craving centered on the within-reach. It is not just the craving that places the addict before the vulnerability of his/her addiction, or else animals would display a similar proclivity. On the contrary, it is the habituation of the craving that creates the problem, or more precisely, the allowance of the compromise which continually redirects the addict to the convenience and the

---

[9] Heidegger 1977/1962, pp. 259–260; tr. 240.

[10] See Zimmerman 1995, p. 503.

comfort of the within-reach, the problem of finding the strength of one's existence based on something else, *dependency*.

Heidegger describes falling as the self's tendency to be "tempted" and "tranquilized" by those concerns which circumvent the difficulty of life's predicaments.[11] And the allowance of the compromise, both in terms of the dependency of the addict and the co-dependency of the acquaintance, defines what we commonly call "*denial*," in short holding to the illusion that, despite all the evidence to the contrary, there really is no problem. Such denial amounts to what Heidegger describes in existential terms as "*evasive turning-away*."[12] The dissimulation inherent in this evasiveness, which in common parlance amounts to deception, excuse-making, and lying (most of all, to oneself), determines the momentum of the downward spiral whereby the self become bound to its craving, enslaved to its fixation, in short, "alienated" from its existence. "The alienation of falling–at once tempting and tranquillizing–leads by its own movement, to Dasein's getting entangled [*verfängt*] in itself. The phenomena we have pointed out—temptation, tranquillizing, alienation, and self-entangling (entanglement)—characterize the kind of being which belongs to falling."[13] If it were not for the self's capacity to fall, and become entangled in the web of its own deceptions, there would not be something like addiction. Because falling, and the concomitant behaviors of lying, deceiving, and excuse making belong uniquely to human existence as care, as we have seen, can individual succumb to their own self-destructive tendencies. In falling, one forsakes one's identity, as unowned or inauthentic, for some illusory substitute, e.g., the histrionics of gambling, which mirrors a desire to recapture the unity of the self that is otherwise lost due to the fragmentation incurred from being lost in the "they." But such attempts to recapture one's identity only amplify the self's fleeing, fugitive tendencies, which fuel the craving(s) of addiction.

As one of the primary existential preconditions of dependency, spatiality points to an important link between "availability," as the impetus to bring "within-reach," and the process of becoming dependent, which is commonly described as "habituation." For an addiction not only becomes possible when an individual fixates on the "within-reach," but also requires the habituation thereof. Along with the spatiality of "de-servering" as a precondition, we must also consider a second existential precondition: that is, a falling mode of temporality, which abandons the self to the relentless pursuit of its cravings. Indeed, the process of becoming habituated, as even the popular language shows with such allusions to certain "substances" becoming "habit-forming," implies a temporal dimension–as what has transpired in the past, but is also experienced in the present and *continued* throughout the future. Yet, while a habit may suggest a "past" behavior, the "future" must also be implied as the avenue, which clears the way for returning to the past, and thereby providing the *opportunity* for the individual "repeating" the same behavior. Here we discover an important clue as to how addiction takes the form of dependency, which in turn is

[11] Heidegger 1977/1962, pp. 235–239; tr. 221–224.

[12] Heidegger 1977/1962, pp. 182–183; tr. 175.

[13] Heidegger 1977/1962; p. 238; tr. 223. Also see Seeburger 1993, pp. 6–19.

lived and acted out as a habituated form of behavior. The acting-out means that the individual returns over and over again to engage in the same behavior, despite the destructive consequences that may ensue.

Let us take a closer look at this form of habituation. Heidegger suggests that habits are developed from the self's facticity, that is, from the preoccupations that come from the self's already existing in the world and through its pre-orientation toward what is immediately available in the environment. Because the individual experiences this facticity on an everyday basis, the preoccupations of life are endemic to the structure of everydayness. Even in the simplest pursuits various preoccupation may take hold, for example, when I am preoccupied before teaching my class with getting to the ATM to extract cash for lunch in the cafeteria afterwards. But once again a temporal dynamic shapes the most basic preoccupations. In the case of addiction, however, the preoccupation becomes all-consuming, usurping in advance all other concerns and interests. When the preoccupation then becomes centered on a specific concern and thereby possibility presented by the environment, e.g., taking a drink, then what the self has done before, e.g., overindulging (in the past), is what one seeks to do now (in the present), because that is also what is achievable (in the future). Thus, as anchoring the dependency of addiction, habits correspond to a specific kind of comportment (*Verhalten*), unfolding within a temporal arc subordinating the future to the past. In this negative form of habituation, the self does not reopen for choice the possibility pertaining to this or this activity or comportment. Instead, the exercise of care is narrowed to the specific gratification of desires and choosing is restricted to instrumental steps requisite (as means to the ends) requisite for satiating the desire.

Merleau-Ponty's distinction between "motor" and "perceptual" or behavioral habits proves to be instructive here.[14] In the latter case, any activity that involves our "handiness" already implies a habit. In a positive way, for example, in the case of a hobby, we might identify a professional golfer's forward press as a trigger to initiate his/her swing. But the habituality of the use of our hands can also take on a negative or destructive character, for example, in the way in which smoking tobacco is as much about what the hands do in lighting and holding the cigarette (as in the satisfaction of nicotine dependency). As the neurobiologist, Erickson, points out: "The 'habit' part of smoking may be related to holding the cigarette, flicking the ashes, or having a cigarette after a meal, with alcohol, or with other drugs. Although the 'habit' is not the chemical dependence, the two combined can make smoking cessation extremely difficult."[15] Habits involve a way of interacting with the environment, in way that takes advantage of the proximity and availability of what can be given, either immediately or even in the narrow window of a short-term possibility, e.g., a future calculated in minutes and hour. For example, in the development of such a destructive habit (*Gewohnheit*)[16] as consuming alcohol, the preoccupation of

[14] Merleau-Ponty 1961, pp. 151–152.

[15] Erickson 2007, p. 133.

[16] For a further discussion of the existential basis of habit, see Heidegger 1988/1999, pp. 97–98; tr. 75. Also see "Translator's Notes," p. 115, n. 29.

thinking ahead to the next drink kicks in as a pursuit, and thereby through a behavioral pattern of indulging over again in that activity. This way of temporalizing also includes a way of spatializing the continuum of means to an end, the desevering by which the individual seeks to bring "within reach" whatever from the environment, e.g., the cocaine or scotch, promising to be a source of gratification.

The advantages of Heidegger's unique description of temporality, as moving on an elliptical path rather than a merely linear one, become apparent through the attempt to address such examples. When looked at in behavioral terms, the graphic characterization of addiction as a way of "chasing the high" betrays a desire to return to a more fulfilled form of life than may be granted either by the present or, for that matter, the future, e.g., seeking to recapture the euphoria, which was achieved the first time one got drunk or played video poker.[17] When looked at existentially, these behavioral tendencies presuppose a distinct way in which the self experiences its time, or temporalizes in its mode of falling. Let us now examine this unowned, fallen mode of temporality, as it defines the unique dynamic of the self's tendency to become addicted through its fixation on the various items which present themselves as within-reach.

In *Being and Time*, Heidegger develops a unique portrait of temporality, which diverges sharply from the traditional, ordinary (i.e., linear) concept of time. He argues that temporality is originally constituted out of the future, not from the emergence, continuance, and passing of the present.[18] Primordial time, as he calls it, emerges from the future, recoils from the past, and then arrives through the present. The future, then, is not some subsequent state we may be passively waiting or hoping for, like winning the lottery, but instead is that in which we, through the development of possibilities, are actively involved in shaping. Conversely, the past is not simply a state which has elapsed and is at best a memory, but rather is that which can be redefined by suggesting ways to develop new possibilities, e.g., the philosophy of Plato learned years ago gives direction for a career change at a later stage in life. Temporality, then, is a dynamic that seeks in the possibilities of the future the creativity to revitalize the past, and hence activate the present, such that the self continually reaffirms the challenge of becoming the unique individual he/she can be.

From a temporal standpoint, however, the case of addiction is quite different. The self resists accepting such a challenge, seeking instead the security and comfortableness offered by the "they." In this regard, Heidegger distinguishes between the authentic, owned self, and the inauthentic, disowned self. Each mode of existence entails a distinct form of temporality, in the former case a dynamic that joins the dimensions of the future and the past in order to order to hold open the significance of the "moment" in which the self dwells. As Heidegger states in *Zollikon Seminars*: "The self is never present-at-hand as a substance. The constancy [*Ständigkeit*] of the self is proper to itself in the sense that the self is always able to come back to itself

---

[17] See Pinsky 2003, p. 143.

[18] Heidegger 1977/1962, pp. 428–430; tr. 372–374. Also see Heidegger 1991/1997, pp. 187–188; tr. 131.

and always find itself still the same in its sojourn [*Aufenthalt*]."[19] But in inauthentic, unowned existence, the self flees from the challenge of individuation, and, in the case of intensified falling, becomes dissociated through its pursuit of various diversions. Accordingly, the inauthentic self temporalizes by substituting one present instance for another, such that novelty is not developed from the future but instead is vicariously relived, as it were, by seeking to recapture by revisiting the past what the present seems to foreclose. The present is not actively lived, but is sacrificed for its supposed discomfort, pain, and inadequacy for an alternative of comfort, pleasure, and fulfillment, which a retreat toward the past supposedly offers. Thus, this fugitive affinity to the past, developed out of intensified falling, re-occurs to the point of fixation, because returning to what is "no longer" can offer the illusion of transposing the self into a refuge of solace. And it is this search for a refuge of solace, which lies at the basis of the common-sense belief that addicts are seeking an escape. But the escape is not only indexed by a "where" (i.e., what is defined environmentally by the set of circumstances yielding the "within-reach"), but, just as fundamentally, by a "when," a fleeing toward the past. *And this fugitive relation to the past, as it were, yields a portal of fleeting escape in the direction of that which, in the form of what is "within-reach," emerges as a source of fixation and an outlet to which one can keeping returning, e.g., the bottle, the joint, or internet pornography.*

Yet what role does the future play in this fugitive process, insofar as the past provides the handle of fixation? The future become a hankering for the constellation of the set of circumstances, e. g., the end of the work-day, which affords the opportunity to "kick back" and retreat vis-á-vis the past, into the refuge of solace. In *Being and Time*, Heidegger designates such a future as unowned, insofar as the self passively "awaits" something to occur by assuming a stance of "expectation."[20] He contrasts the authentic or owned future, in which the self is actively involved in transforming its situation by "anticipating" or reaching ahead to meet its possibilities.[21] As the temporal corollary of addiction, expectation is waiting for something better to come along, a momentary release from one's circumstances, which only the immediate gratification of the "within-reach" can provide. Although the past provides an avenue of escape, the future serves to reinforce the passivity by which the self succumbs *over and over again* to its cravings, i.e., becomes "fixated." When the addict is not literally "escaping into the bottle," the bondage of addiction still holds in the individual futural preoccupation with drinking even while he/she is at work and is prohibited from drinking. As Seeburger properly summarizes this aspect: "...as members of Alcoholics Anonymous often remark, regardless of how much time they actually spent drinking, back in their drinking days, their concern with alcohol was pervasive. As they like to put, even when they were not actually, they were still always 'thinking drinking.' That is, they were planning how they were going to get their next drink...."[22]

---

[19] Heidegger 2001, p. 175. Pearl 2013, pp. 56–60.

[20] GA 2, pp. 386–387; tr. 306.

[21] GA 2, 386–387; tr. 306.

[22] Seeburger 1993, p. 84. For further discussion, see Kemp 2009, pp. 1–18.

When occurring either authentically or inauthentically, temporality always includes three dimensions: future, past, and present. In its inauthentic, unowned occurrence pertaining to addiction, we can distinguish the schemata for the interplay of the three temporal dimensions or ecstases.[23] Adopting a Kantian term, which Heidegger also employs,[24] we can identify the schema of the future as the prospect or "plan" of seeking in what awaits one something better, a passive condition of wishing for a change, no matter how transitory or superficial, e.g., an alteration of mood (relaxation, euphoria). The schema of the present involves the logistics of desevering and making available the expectancy held by the promise of the "within-reach." Within the schema of *making-present*, the self, through the fixity of its addiction, seeks to insure a sense of *stability and continuity*–otherwise lost in the dispersion of its falling–in short, a measure of *control,* over a life which appears directionless and futile. The common sense belief that addiction is a form of immediate gratification holds, but only to the extent that this immediacy is a way of reclaiming control "here and now," e.g., the "rush" of feeling better, and hence implies the past. The schema of the past is the postponing of, or allowing to elapse, the challenge of existence, in exchange for re-treating into a comfort zone of security and solace. Within the addictive experience, each of these three temporal ecstases is already in play and related to one another. Nevertheless, because the mark of addiction lies in its fixating tendency, the past dominates in the sense of falsely subordinating to it the expectancy of the future. Two key points arise. First, the future remains closed-off, because in its expectancy it is already dominated by a previous standard of satisfaction supplied by some past experience (e.g., the euphoria of internet gambling). Second, without the fugitive, escapist character of the past, the obsessive-compulsiveness inherent in addiction would be lacking. Indeed, addiction is like a faulty reconstruction of the routine of everyday life, in which its regimen is not defined by the entire course of one's day but instead by a single activity–e.g., drinking or gambling. As Seeburger remarks in identifying the similar addictive effects that stimulants and depressants may have, despite their contrary medical aims: "Both provide a way of regulating the organism's level of excitation, keeping it constant. It is precisely such sameness, such routine repeatability, that the addict seeks in the object of addiction."[25]

Addiction closes off the future, restricts it to the next opportunity to fixate on a source of immediate gratification, in short, to "get a fix." According to Heidegger, the temporalizing of the future yields an expanse of possibilities, and spawns the fundamental openness of existence. Conversely, by closing off the future, the fixity of addiction constricts the original "ontological openness." Herein lies the ontological "fact" as to why it is so difficult for the addict to break the cycle of addiction and start on the path of recovery. Because deception and denial fuel addiction, the addict can overcome these tendencies only by cultivating the openness of existence which, however, remains most foreign to him/her. Put another way, the impetus toward

[23] See Heidegger, "Die Zeitlichkeit des Verfallens," 1977/1962, pp. 458–461; tr. 396–400.

[24] Heidegger 1977/1962; p. 483; tr. 356–357.

[25] Seeburger 1993, p. 127.

responsibility, to be answerable, is what the addict possesses least, but requires most, if recovery is to be possible. Because of its obvious importance, I will address the question of recovery in a subsequent chapter.

2.4 Conclusion

By showing how everyday modes of spatiality and temporality underlie the self's tendency to be dominated by its cravings and fixations, I have taken the initial step toward outlining the dynamics of addiction. In the following chapter, I will examine how these dynamics extend the "hook" of addiction, the existential preconditions behind the self's "obsessive-compulsive" tendency to fixate on a specific source of arousal.

References

Aho, James, and Kevin Aho. 2008. *Body matters: A phenomenology of sickness, disease, and illness*. Lanham: Lexington Books.

Emad, Parvis, and Kenneth Maly. 1999. Translators' foreword to Martin Heidegger, *Contributions to philosophy (From Enowning)*. Bloomington: Indiana University Press.

Erickson, Carlton K. 2007. *The science of addiction: From neurobiology to treatment*. New York: W. W. Norton & Company.

Heidegger, Martin. 1962. *Being and time*, trans. John Macquarrie and Edward Robinson. New York: Harper & Row.

———. 1977. *Sein und Zeit, GA 2*. Vittorio Klostermann: Frankfurt am Main.

———. 1983. *Die grundbegriffe der metaphysik*, GA 29/30. Frankfurt am Main: Vittorio Klostermann.

———. 1988. *Ontologie (Hermeneutik der faktizität)*, GA 63. Frankfurt am Main: Vittorio Klostermann.

———. 1991. *Kant und das problem der metaphysik*, GA 3. Frankfurt am Main: Vittorio Klostermann.

———. 1995. *The fundamental concepts of metaphysics*, trans. Nicholas Walker and William McNeill. Bloomington: Indiana University Press.

———. 1997. *Kant and the problem of metaphysics*, trans. Richard Taft. Bloomington: Indiana University Press.

———. 1999. *Ontology—The hermeneutics of facticity*, trans. John van Buren. Bloomington: Indiana University Press.

———. 2001. *Zollikon seminars*, trans. F. Mayr and R. Askay. Evanston: Northwestern University Press.

Kemp, Ryan. 2009. The temporal dimension of addiction. *Journal of Phenomenological Psychology* 40 (1): 1–18.

Merleau-Ponty, Maurice. 1961. *Phenomenology of perception*, trans. Colin Smith. London: Routledge,

Pearl, Joel. 2013. *A question of time: Freud in light of Heidegger's temporality*. Amsterdam: Rodopi.

Pinsky, Drew. 2003. *Cracked: Putting broken lives together again*. New York: Harper-Collins.

Seeburger, Francis. 1993. *Addiction and responsibility*. New York: Crossroads.

Zimmerman, Michael E. 1995. Ontical craving versus ontological desire. In *From phenomenology to thought, errancy, and desire: Essays in honor of William J. Richardson, S. J*, ed. Babette Babich. Dordrecht: Kluwer Academic Publishers.

Chapter 3
The Phenomenon of the Body and the "Hook" of Addiction

In the previous chapter, we uncovered the existential preconditions for "dependency" occurring. But there is another side to the problem. Besides the question of how the self can become addicted, it is equally important to address the "captivating" character of any addiction, its distinctive force of arousal and the *fracturing of desire*, which may translate into different kinds and combinations– i.e., the phenomenon of "multiple addictions." Put in more graphic terms, we must consider the *experience* of the addiction from "inside," as it were, its power and efficacy or what we might call the "hook" of addiction, the abysmal source of its cravings.

How do people become "hooked?" The everyday locution already serves as an "indicator" of the phenomenon of addiction, that is, its existential locus. While Bill Wilson was an important historical figure in the twentieth century, his life also stands as a metaphor for the "everyday" occurrence of addiction. Alongside dependency, we must also consider a facet that Wilson's cultural situation also illustrates, i.e., as a stockbroker who stood on the cusp of what America would become as the land of "conspicuous consumption." For the flip side of dependency is the insatiable thirst to consume whether it is a "substance" like alcohol, as we will see, some other product like internet pornography. As Jane Valez-Mitchell has illustrated in a more popular forum, in the United States particularly—as the wealthiest nation in the world—we have witnessed the crossover between the materialistic drive of consumption and the rise of addiction to an epidemic level.[1] We have also seen instances of a changing topography in which the pursuit of habit-forming synthetic opiates has led those of the most affluent means to seek avenues initially traversed by the poorest members of society for obtaining illegal drugs like heroin (to quell their addictions).[2] For example, as recently as the summer (2017), the governor of Louisiana has declared opiate addiction to be a problem state-wide.

[1] See Mitchell and Mohr 2011, pp. 19–23.

[2] For one of the most graphic depictions of this problem, see Quinones 2015 ("Preface").

F. Schalow, *Toward a Phenomenology of Addiction: Embodiment, Technology, Transcendence*, Contributions To Phenomenology 93,
DOI 10.1007/978-3-319-66942-7_3

In emphasizing that there are different forms of addiction, we arrive at an important question. How can addictions seems to divide into two distinct kinds, the one more obvious involving dependency on a specific substance, whether it be drugs or alcohol, and the other less as involving over indulgence of an activity or "process," whether it be gambling or surfing the Internet. The amorphous character of addiction, including its overlap with others kinds of obsessive-compulsive disorders, sexual deviancy, and distorted body-image, dissociative personality, indicates why addiction is so difficult to define. Despite many attempts to circumscribe an "addictive personality," the search for a "nature" or "essence" of addiction remains something of a chimera. To employ Ludwig Wittgenstein's term, we might be better served to speak of "family resemblances" to describe the polymorphic character of the phenomenon, if it were not for the ambiguity that in fact there are "familial," "ancestral," and "generational" ties in the personal history of many addicts.

In order to address this elusive aspect of addiction, we must continue to unravel the problem on a even deeper level of inquiry, that is, which re-examines the self through the unique circumstances of its "embodiment." The simple *assumption* that there is a "mental" as well as a "physical" side to addiction, and that this distinction is mirrored in the way that we split the problem in two sides (i.e., as either "process" or "substance"), spearheads further inquiry. If hermeneutics aims to clarify such presuppositions, then we must proceed to do so by showing how the "lived body" precedes the split between "mental" and "physical," and thereby points to the deeper origin of our desires. Yet if this origin is not merely to be merely conjectured, we must distinguish a corresponding phenomenon, an example, however primitive, to illustrate a locus of arousal that underlies any addiction.

In this chapter, I will, first, point to the formation of a *fetish,* in order to show how desires can be centered and ultimately "fixated" on specific points of attraction. In this way, I will outline the landscape in which the things and activities which punctuate the "world" can emerge as "lures of attraction," the spinning of the "hook" for any addiction. Accordingly, we will come to appreciate more fully a primordial, if not primitive root of addiction that indicates why its "hold" is so difficult to break once the "hook" has been set. Secondly, I will take a detour around the problem of addiction as ordinarily understand, and consider this primordial basis of "desire," as implicating the circumstances of our embodiment, on the one hand, and providing a foothold for the development of fetishes, on the other. Thirdly, I will then establish the origin of what might be called "indiscriminate attraction," that is, the investiture of what is "within-reach" in our environment to disproportionate significance, the hallmark of a fetish. Fourthly, I will show how, within our culture of conspicuous consumption, a fetish becomes an icon to explain the "hook" of addiction, a "hyper extension" of desire, as it were, which traverses the divide between substance and process or behavioral addictions, and thereby points to *the phenomenological basis of a distinctive pathology that is culturally as well as individually based*. Given this hyperbolic extension of desire, we will discover how, to transpose Karl Marx's famous aphorism, "opium becomes the 'opium of the people.'"

3.1 Image, Gesture, and Desire

In re-examining the dynamic of human desire, we cannot overlook a distinctive capacity that fuels and intensifies desire(s), and that is the power of fantasy. While it is possible to equate desires with instincts, the character of our embodiment suggests a more complex nexus, which also includes the "gestures" expressed in the desires: the semiotic and symbolic dimensions that speak through the images of fantasies. The fact that our desires are not simply reducible to physical impulses, but instead spiral out into diverse icons of fantasizes, suggest that (1) the fact of "being-embodied" corresponds to the self's thrownness into its situation[3] and, conversely (2) that the self's *incarnation* is dynamic, that is, occurs through the "ecstasy" of its transcendence, i.e., as projecting open a horizon of possibilities through its being-in-the-world.[4]

However we may describe the link between desire and fantasy, an important facet nevertheless emerges in regards to the problem of addiction. For despite all the "sensual" character of one's indulgences, much of their impetus and intensity depends upon their "figurative associations," or what we might simply call the "play of fantasy." The so-called "compulsiveness" of an addiction would not have its power without the accompanying "obsessiveness," that is, the images, associations, and fantasies that enhance our desires and allow it to *monopolize* one's concerns. Though through its production of fantasies, what Heidegger, following Kant,[5] calls "imagination" (*Einbildungskraft*) may be associated with a mentalistic function, we will discover that, quite the opposite, its manner of disclosedness depends upon our embodiment. Conversely, the development of the imagination on this concrete footing provides us with a springboard to overcome the metaphysical dichotomy between the mind and the body, while uncovering the latter as a lived dimension of our being-in-the-world. The foremost instance of this lived dimension, as exemplifying the transmutation of desire into fantasy, is, of course, the "play" of our sexuality.

The emergence of sexuality as a phenomenon implies this bodily dimension. On the one hand, imagination surpasses what is merely given within an environmental context, and hence has as its target something "hyperphysical." On the other hand, whatever its innovativeness, the sexual character of the fantasy cannot completely detach itself from its tie to an environmental context, and hence reinstates a "reference" to the physical, albeit in a transformed and inverted way. In the guise of a "fetish," the physical re-emerges as the creative wellspring for fantasy, as constellating the concrete point or locus of arousal. The vestige of embodiment within the sexual fantasy, then, comes to light in the corollary need for the development of fetishes. To be sure, the task of psychology is to address the dangers that are associated with an individual's becoming fixated on a specific sexual fetish, e.g., items of

[3] See Schalow 2006, pp. 7–17.

[4] Schalow 2006, pp. 1–9.

[5] Heidegger 1991/1997, p. 198; tr. 138.

clothing, and, subsequently, the development of a specific pathology, i.e., sexual addiction. In the case of sexual addiction, bodily desires are channeled into the narrowest avenues of arousal, thereby converting the self's erotic impulses into a longing for immediate gratification.[6] The longing is not simply an internal obsession, but instead becomes represented environmentally as a specific item to help consummate the arousal, in other words, by acting out one's desires.

In terms of the scope of this phenomenological analysis, my aim is to describe a fetish as a phenomenon that illustrates the emergence of imagination within the situated context of our embodied existence. Yet, in the course of addressing the phenomenon of sexuality, it becomes necessary to discern the ambiguous character of fetishes. On the one hand, fetishes provide a source of stimulation to supplement fantasies; on the other hand, because of their tangibility, they can usurp the power of fantasies and thereby induce a "fixation" on specific physical artifacts. By completely "closing off" the "play-space" of erotic imagination, such a fixation rises to the level of an addiction. While Heidegger may not have explicitly addressed the erotic imagination, his discussions of the creative power of imagination are specifically oriented toward a pre-predicative dimension of meaning. The hermeneutic orientation of his phenomenology allows for a more explicit account imagination than may accessed on the basis of Kant's transcendental precepts alone.[7] By contrast, in Kant's case, the imagination initially surfaces in an epistemic role to generate the schemata for the application of the pure concepts of understanding to objects of possible experience. Depending primarily upon a model of predication, Kant considers the capacity of the pure concepts to "refer to" or signify the determinate features of an object, in conformity with the condition of our finitude as knowers. The actual dynamics by which the categories acquire their content, and hence the source of their significance, becomes secondary to their predicative role in the articulation of the universal principles of human knowledge and ultimately the laws of nature. In his destructive-retrieval of Kant's thought, however, Heidegger argues that the imagination's role in schematizing the pure concepts, and in yielding their content through "transcendental determinations of time,"[8] suggests another level of the genesis meaning prior to the act of predication.

If we consider imagination as one axis of creativity for producing images and perhaps fantasies, then intersecting with it we can postulate another axis of language or discursivity. For Kant, this intersection is shaky at best, because for him the problem of meaning remains implicit within the confines of the first *Critique,* only to emerge tangentially in his account of reflective judgments of taste in the *Critique of Judgment.*[9] Heidegger, on the other hand, has an easier time marking this interface, because for him the pre-predicative level that his hermeneutics uncovers suggests an affinity between the creative production of images and their capacity for signification, and ultimately expression in gestures, figures, and words. In terms of

[6] See Kemp 2009, pp. 1–18.

[7] Heidegger 1991/1997, pp. 134–141; tr. 94–99. Also see Schalow 2016, pp. 377–394.

[8] Kant 1965, A 139 / B 178.

[9] Kant 1951, pp. 23–29. Also see Schalow 2017, pp. 213–229 (esp. 222–226).

his fundamental ontology, Heidegger describes this crucial linchpin, which brings forth the centrality of imagination as defining our temporal constitution qua finite beings, a mode of "disclosednesss" (*Erschlossenheit*). Thus, imagination constitutes an original act of disclosednesss, whose dynamic becomes most pronounced in the way that language opens up a world of meaningful involvements, possibilities, and pursuits.

This openness in which we participate by speaking, however, develops a wide range of meanings that extend from a preverbal level of "gestures" to a thematic level of concepts. Through his hermeneutics, Heidegger recognizes that gestures can elicit patterns of meaning, and, within the overarching context of signification of the world, serve as vehicles of communication. At first the gestures may assume only a particularly innocuous form, such as a casual wave of salutation or greeting. The fact that the pre-verbal gestures may not imply anything sexual, however, does not mean that in another context they could expressly do so. The case in point might be a "wink," that in one context could be interpreted as primarily salutatory, and in another context might suggest an overtly sexual attraction between two people.

Temporarily, we need to defer this concern as to whether Heidegger addressed sexual concerns, in order to explore issue that he did consider, albeit not as centrally as other thinkers like Merleau-Ponty,[10] namely, the correlation between the use of gestures–with their power of signification–and the character of our embodiment or incarnality. Even within the mundane, equipmental context of the ready-to-hand, Dasein's dexterity in using tools, and the significance of doing so in developing the individual's possibility, e.g., building a house, the condition of embodiment accompanies of the fact of our being-in-the-world. Given this situated network of involvement and signification, our hands are not simply physical appendages, but also include the power of signification. As Heidegger states in *What Is Called Thinking?*:

> The hand does not only grasp and catch, or push and pull. The hand reaches and extends, receives and welcomes–and not just things: the hand extends itself, and receives its own welcomes in the hands of others. The hand holds. The hand carries. The hand designs and signs, presumably because man is a sign. Two hands fold into one, a gesture meant to carry man into the great oneness.[11]

Heidegger's tantalizing suggestion that "man is a sign," and hands belong to humanity's signifying power, goes a long way toward explaining how we experience meaning outside the framework of predication. Conversely, Heidegger argues that even the so-called formalization of language in writing harbors a vestige of a hand and recalls the condition of our embodiment as situated, finite beings:

> The essential correlation of the hand and the word as the essential distinguishing mark of man is revealed in the fact that the hand indicates and by indicating disclosed what was concealed, and thereby marks off, and while marking off forms the indicating marks into formations [*indem sie zeigt und zeigen zeichnet und zeichnend die zeigenden Zeichen zu*

[10] Merleau-Ponty 1961, pp. 174–197.

[11] Heidegger 1968, p. 9.

> *Gebilden bildet*]....The word indicated by the hand and appearing in such marking is writing. We still call the theory of the structure of language "grammar."[12]

The reflexivity of the fact that we ourselves our "signifiers," and can communicate thereby, implies we can also address an aspect central to our embodiment, which directly or indirectly pertains to the character of our "thrownness" into the world. Heidegger suggests that human existence is factically dispersed in its thrownness, in such a way as to allow for the differences between the genders. And regardless of the uncontrollable "fact" of whether the individual is born as either a female or a male, each individual shows concern for, and struggles with, the ineluctable mystery, enigma, and problem posed by his/her sexuality. Insofar as Dasein is essentially "being-with" others, the self's struggle with its sexually includes the various complications of the relationships spawned thereby, from dating to parenting, from marriage to divorce. And, by the same token, that struggle contributes to the development of the self's identity through the "intertwining nexus of relationships" between it and its world that "forms the medium of intercorporeality."[13]

Beginning with Derrida, various critics have argued that Heidegger ignores the importance of the "sexual difference," of "gender," as a crucial dimension of our humanity.[14] Despite the validity of these criticisms, the conclusion should not be drawn that Heidegger denies that sexuality is germane to human existence, or that it cannot it factor in as a primary concern to "formally indicate" the enactment of care (*Sorge*) as such.[15] Indeed, the formally indicative character of human existence implies that the individual's way of confronting his/her embodiment harbors a latent message, as it were, as to the relevance that his/her needs and desires may have. Insofar as Dasein is not simply an impersonal "what," but also a "who," the most rudimentary affects and responses can speak to the inescapable facticity of human existence. Because through the temporalized act of existing the individual addresses who it is in a meaningful way, even the most primitive needs and desires of sexuality spawn "figures" of meaning or signification. And since Heidegger does not limit language to predication, but instead traces its origin back to a "pre-predicative" level of experience,[16] he can address the hermeneutic import of the random, idiosyncratic, and "impulsive" aspects of our sexuality.

Because, on the one hand, the issues of an individual's sexuality underscore its embodiment, and, on the other, point to the uniqueness of who each of is, their manner of expression is never simply univocal. Rather, in struggling to come to terms with, address, and express our sexuality, we experience a fissure within ourselves, of the physical and the emotional, of desire and devotion, catapulting us into an openness that unfolds on two fronts. Because tension and ambiguity are the hallmark

---

[12] Heidegger 1982/1992, p.125; tr. 84.

[13] Here Maurice Merleau-Ponty's discussion of embodiment becomes instructive. See Steeves 2004, p. 119.

[14] Derrida 2001, pp. 53–72.

[15] For a discussion of the importance of gender in this regard, see K. Aho 2007, pp. 137–155.

[16] GA 29/30, p. 496; tr. 342.

of our sexuality (crossing the gap between our embodiment and the meaning it has for us), its manner of expression necessarily lends itself to suggestive figures and images. If we trace the source of these images back to the power of imagination, and ascribe to them an explicit sexual content, then we can refer to fantasies to describe the erotic play of this imagery.

We have now shown how imagination can provide a setting for the play of erotic fantasies. We have thereby shown that there is a link between imagination and our embodiment. In order to clarify this link, let us consider the converse: namely, how our sexuality can inscribe itself in facets of our embodiment that are specifically suited for playing out sexual fantasies, i.e., via "fetishes" or, put another way, how our embodiment provides the *site* for erotic play.

3.2 The "Play-Space" of Fantasy and the Lived-Body

Within our psychoanalytic culture, there is a sharp tendency to equate fetishes with a specific sexual behavior. While there is an erotic element within fetishes, if by that we mean a locus of "attraction," they would not be possible without a semantic axis which allows for the "layering" if not "doubling" of meaning. For a fetish arises when something of a physical origin, i.e., an article of clothing, can signify an additional meaning beyond that, i.e., a further locus of attraction, excitement, or arousal, other than what is "literally" represented by the thing's primary purpose. Most notably, Karl Marx recognized that there can be "economic" fetishes, in which the context of signification proper to capitalism and its 'world,'[17] e.g., something providing transportation in the immediate environment like a car, can "signify" an "added allure," an incremental "value" beyond its simple function of use (hence its fetish-like character), e.g., social status, even sex appeal, if that automobile is a Jaguar convertible. The example from Marx proves instructive, because it suggests that any fetish must arise within an overall context of signification, whose unity depends on a prior act of (Dasein's) projecting-open a world. Karl Marx reserves the term "fetish of the commodity" to describe the "iconic" character of whatever fuels the self's desire within the culture of conspicuous consumption. In Heidegger's phenomenology, the world as an encompassing structure of significance surpasses any ontical context of significance, for example, the 'world' of the capitalist within which the "fetish of the commodity" first arises.

When we contrast Heidegger's description of the "world-openness" of human beings, from the fact that animals are "world-poor,"[18] we arrive at the insight that illuminates the key dimension of fetishes. Are animals capable of having fetishes? Indeed, the history of philosophy casts a long shadow in underestimating the capability of animals. Be that as it may, the point here is that only through the prior opening of a world can a range of significations arise, which supports multiple

[17] Marx 1963, pp. 216–217. Harrington 1967, p. 171.

[18] Heidegger 1983/1995, pp. 509–511; tr. 350–351.

levels of meaning as well as allows for ambiguity. As long as the ambiguity of meaning is possible, and there is a context of embodiment to unfold that possibility, we can safely say that something on the order of fetishes are possible. Indeed, a fetish *becomes possible* by referring to a second level of meaning, whose first level remains rooted in the lived dimension of our embodied being-in-the-world, e.g., the nexus of equipmental relations. Given this phenomenological root, we can thereby define a fetish as the endowment of a power to "attract," thereby adding a further level of signification, to what that same thing may ordinarily have within its original context of familiarity. Thus, to take a simple example, black nylon stockings may provide an additional source of arousal, beyond the simple use-value a pair of stockings may have in clothing a woman's legs within its familiar context of signification. In more extreme cases, the fetish extends the range of "imitation" or "*mimesis*" to include "cross gender" activities, e.g., wearing the clothes of the opposite sex or transvestism. In this context, we need to clarify (1) that the fetish has a signifying element to it, what we might call an "iconography," in connection with its tie to our embodiment; and (2) in this regard, we can characterize the fetish as a "physical icon," in contrast to the "image" of a fantasy, which displays the creativity of imagination and is a conduit for its "play." Thus "play" or innovation is essential to the "iconography" of the icon, or, put another way, the latter's power hinges on its "polyvalency," i.e., or the ability to spark arousal by exploiting the ambiguity of meaning(s).

While the concern for fetishes was obviously not central to Heidegger, any more than sexuality was, he nevertheless makes a few casual references to them in his discussion of "world" in the first division of *Being and Time*. He singles out a fetish (*Fetisch*) as an instance of Dasein's absorption in "everyday concern," particularly as exemplified in so-called "primitive" peoples ("*primitiven Dasein*").[19] But even when the individual is preoccupied in this way, and seeks its identity through the "serviceability" of something in the environment, signs guide this mode of concern. Indeed, the sign is the intermediary between the concern that preoccupies the individual and whatever emerges within the environment as the source of the fetish. Thus, there might be some kind of emblem or insignia that distinguishes the fetish. In so-called "primitivism," or even today when sexual activities border on the bizarre, fetishes may be accompanied by certain kinds of rituals or repeated ceremonies commemorating an event, e.g., as in the re-enactment of various sadomasochistic scenarios.

The pre-theoretical use of signs in fetishes suggests that there is a close proximity to whatever emerges within the environment as a potential source of eroticism. But, conversely, the signifying acts occurring a fetish presuppose the wider context of significance of the world, which discloses not only what is immediately encountered in the environment as the specific source of arousal–the thing, equipment, ready-to-hand article—but also the "*eros*" of desiring itself as a unique concern of Dasein or what Heidegger calls the "for the sake of". Without the structure of "that for the sake of" there could be no fetishes, since it is a mistake to equate the fetish

[19] Heidegger 1977/1962, pp. 108–109; tr. 112–113.

simply with a physical object or worldly referent. Indeed, "that for the sake of," in referring equally to a possibility of Dasein as well as the specific thing's emergence within a context of significance, distinguishes the "space of engagement" whereby the self can direct its attention to a "second" meaning (of arousal) above and beyond the "first" meaning of the thing's primary function of use (as an item of equipment). Without the "allocation of this space," which unfolds along the wider arc of the world including "that for the sake of," as well as the "in order to" of the equipment, a leeway for "disclosure" would not be created in which a fetish could arise. As Heidegger emphasizes in discussing "mythic thinking" in his critique of Ernst Cassirer's work (1925), the dynamic of spatiality allows human beings to comport themselves toward things in the environment with an attitude of fascination and awe, through ceremony, ritual, or even (sexual) rights of passage.[20] But without space giving direction to such behaviors, of proximity and distance, of intimacy and belonging, neither could we relation to things with the appropriate response nor could they appear with a pregnancy of significance.[21]

Though Heidegger employs this term more broadly, we can call this room for disclosure the "*Spiel-Raum*" or "play-space." By occupying this play-space, Dasein can cultivate a multiplicity of possibilities and thereby disclose itself as an inhabitant of the world. Because it occurs in tandem with the world, the play-space that Dasein inhabits is never "free-floating." This is particularly true when we consider its importance in the arena of eroticism and fetishes. For this arena is specifically outlined by the openness we experience due to our embodiment. Within this openness Dasein defines itself through a form of "directionality," of going "toward" and "away," an "origin" and a "beyond." If these "vectors" give expanse and depth to Dasein's openness, then the self, through its embodiment, is already drawn into this arena in which it can engage in the "play" of eroticism. If only euphemistically, we can call this arena an "erotic zone." Be that as it may, this zone is comprised not by any specific set of acts, but by the possibilities that emerge in the course of being-in-the-world as such.

However we postulate this arena of erotic play, it must already be "pre-cleared" in terms of the existential constitution of the "there," or Dasein's disclosedness. Indeed, a so-called "erotic zone" is not something additional, but instead emerges as a phenomenon we can consider only by presupposing the disclosedness of being-in-the-world and the unique dynamic of this "play-space." When Heidegger outlines the existential constitution of the "there," and identifies its three structures of dispositions, understanding, and language, he emphasizes its "ecstatic" character. The irony of his characterization is that, despite its ontological meaning, it also harbors erotic connotations. Yet for Heidegger the "carried-awayness" of the temporal "ecstases" yields the "directionality" of all the self's other concerns and interests, simply because temporality emerges as the "ontological meaning of care." If Dasein

[20] Heidegger 1983/1995 pp. 268–269; tr. 188–189.

[21] For a discussion of the importance of space as an element of Da-sein's embodiment, see C. Cicoan 2008: 80. For an excellent account of Heidegger's disinterest in addressing the problem of embodiment, see S. Overgaard 2004, pp. 116–131.

cultivates specific possibilities, which exhibit a similar trajectory, then they arise only on the basis of a prior, ecstatic projection of its own potentiality. As a result, only given this prior clearing can the individual, from the exalted heights and deepest roots of his/her embodiment, experience the frenzied excitement of sexual release. The telling point of the priority of Dasein's ecstatic nature, as ontologically conceived, is that the unique dynamic of this "play-space" lies in "making room" for the further amplification of the sexual experience–increasing its intensity and extending its longevity.

The ecstatic character of the "play-space" makes it possible for a "doubling," as it were, of the individual's concern for his/her sexuality. What presumably begins as an interest in, and a response to, a physical desire, transposes itself into an openness of possibility that amplifies its power. Because this openness cannot be simply reduced to the merely corporeal, its emergence necessarily includes the entire structure of significance comprising Dasein's being-in-the-world. When read off this context of significance as a whole, specific signs or "icons" may emerge which are endemic to this sexual play, and, in the guise of fetishes, etc., contribute to its amplification. The ecstasy of the opening of Da-sein's "there" assumes an additional meaning, as implying the self's ascent to the peak of eroticism.

We have now shown how Heidegger's fundamental ontology, his account of the ecstatic constitution of the "there," paves the way for considering a range of sexual concerns, including their potential signification through such facets of our embodiment as "fetishes." Having established our how embodiment provides the "site" for the play of eroticism, let us consider a specific expression thereof through an enactment of the "play-space" that produces fantasies or sexual images, the "imagination."

3.3 Fetishes and Their Unique Pathology

While imagination is often construed as epitomizing the flight of fancy, we cannot overlook that its context of signification still betrays a worldly origin. In the *Kant and the Problem of Metaphysics*, Heidegger establishes the link between Dasein's self-disclosive capacity and the power of imagination, which, as the harbinger of ecstatic temporality, creates the "play-space" for the manifestation of what is. And, insofar as temporality is the hallmark of human finitude, imagination is a power that human beings exhibit only because they are finite. Moreover, finitude is not simply a negative characterization, but as the premise of setting limits already includes an impetus to surpass toward them. Eroticism is a capacity ascribed to human beings, not simply on the physical basis of sexuality, but, because the trajectory of our finite transcendence already aims at the furthest vistas of our limitations—a point that Giorgio Agamben (who attended Heidegger's "Seminar in Le Thor 1966") makes: "Lovers go to the limit of the improper in a mad and demonic promiscuity; they dwell in carnality and amorous discourse, in forever-new regions of impropriety and facticity, to the point of revealing their essential abyss."[22]

Conversely, in signifying this ascent toward these vistas, imagination maintains a prior orientation or reference to our embodiment, to the thrownness that differentiates each of us by gender and roots the individual in his/her incarnality.[23] If world marks the overarching structure of signification, then it is by eliciting further valences of meaning from the basic gestures of the body that the erotic imagination is born. This polyvalency (of meaning) allows a single phenomenon to implicate multiple levels of significance, and thereby gives fuel to the creativity of imagination.[24] In this way, the power of imagination is unleashed to develop the greatest possibilities of fancy, for the significations that it can draw upon are as vast as the entirety of human experience itself, i.e., the world. In the erotic imagination, what distinguishes the manner of the individual's engagement with it, also distinguishes its manner of disclosedness, the "play" of the "play-space." Though play is often thought of as effortless, this does mean there are not possibilities for its enhancement. Let us take a specific sexual activity, which above all requires fantasy, and bears the namesake in question, or "role-playing." The fantasy and its manner of "acting it" occur in tandem, such that the specific possibilities enlisted not only have a "worldly" reference, but, because at least the participation of one other individual is involved, emerges within the situated domain of our being-with others. Here the erotic imagination calls upon a dynamic central to the craft of art itself, or "*mimesis*." For the fantasy is simultaneously an "imitation" of the sexual act, but, through the vision of the erotic imagination, re-crafted, as it were, within the wider expanse of the "play-space" at the heart of "role-playing." This *mimesis* thereby illustrates how the erotic imagination transposes the physical act of sex into a signifying gesture, allowing for its (i.e., one's sexuality) emergence within the wider expanse of possibility.

Insofar as temporality constitutes human existence as care, and care as the being of Dasein is implied when it expresses its sexuality,[25] a given sexual fantasy necessarily replicates this temporal character. Thus, the erotic imagination not only pieces together a graphic scenario of sexual possibilities, but crafts them in order to juxtapose what withdraws into absence as a farfetched wish with what simultaneously has becomes present as the figures of people, places, and things punctuating one's environment. The mimetic act, then, which sparks the fantasy, then, come to fruition as the innovativeness, which ignites the creativity of the erotic imagination. In a manner which is doubly consistent with Heidegger's view of the central role that temporality plays in both the constitution of human existence and the genesis of the imagination,[26] the temporal dynamic of absence/presence lies at the root of all erotic fantasies. And animating that dynamic is the trajectory of "anticipation"

[22] Agamben 1999, p. 204.

[23] Heidegger 1978/1984, p. 173; tr. 137.

[24] For further discussion, see Boss 1949, pp. 43–44.

[25] Heidegger 2001, pp., 80–81, 139, 200.

[26] Dillon 1993, pp. 316–325. Also see Schalow 2006, pp. 60–64.

(*Antizipation*), which projects the vector of possibility *as* possibility. In regard to its temporal constitution, this vector defines the "directionality" or "*worauf*" of the erotic fantasy, or, in the language of Kant's account of imagination, the "schema" of the future. In simplest terms, the erotic craft of *mimesis* takes shape through this temporal schema, underwriting the "meaning" in the play of fantasy–whether the discrete images are intertwined to form a "narrative" or evenly loosed grouped to constitute a vortext of eroticism and arousal.

Though Heidegger not does explicitly address sexual fantasy, his allusion to "mythical fantasy," in his critique of Ernst Cassirer, proves to be illuminating. Specifically, Heidegger remarks that Cassirer overlooks the "labyrinthine and abysmal" side of fantasies, which can only be uncovered when "guided by the phenomenon of the transcendental power of imagination, and its ontological function within the *Critique of Pure Reason* and the *Critique of Judgment*."[27] Insofar as sexual fantasies in particular may have this subterranean character, their significance depends upon "configuring" them in the "images" of creativity that spring from the imagination. Given this wellspring of creativity, the key to the erotic imagination lies in its novelty, that is, as a medium that is always in the process of reinventing itself and giving new expression to its creativity through the power of "play." In this regard, the erotic imagination becomes a point of departure, as "lover's leap," that targets, in Agamben's word, the "limit of the improper" or what is taboo and forbidden.

In fostering this play, and thereby increasing this efficacy, fetishes assume a special role. Indeed, there needs to be some kind of sensual point of reference presented from the "environment," whether visual, auditory, or perhaps even olfactory, which emerge to "fill out" or "concretize" the significance of the fantasy. As today's Internet, a technological medium for dispersing sexual images attests, the spectrum of fetishes seems to expand into every nook and cranny of the erotic imagination, even into its darkest corners. And, as in every aspect of human concern, there can be another side that can be as entrapping as liberating. In ontological terms, the "openness" has an elasticity to it, which can always snap back into its opposite, or closing off. And the erotic imagination, at least to the extent that it seeks tangible examples to fill out its images, is no different. For the continual enumeration of fetishes may lead down a dark path which gives priority to the examples themselves, and sacrifices the openness of erotic imagination for narrower and even truncated expressions thereof. Though it would take us too far afield here, Heidegger's confidant and friend, Medard Boss, outlined the roots of such "fixations," and their destructive potential to the individual's identity, in his book *Meaning and Content of Sexual Perversions*.[28]

In revisiting the importance of his critical exchange (*Auseinandersetzung*) with Kant in *Contributions to Philosophy*, Heidegger makes this provocative statement: "Imagination" as occurrence of the *clearing* itself.[29] In this same context, he emphasizes that the meaning of the "to be" clears itself only in tension with its opposite

[27] Heidegger 1991/1997, p. 269; tr. 189.

[28] Boss 1949, pp. 145–146.

[29] Heidegger 1989/1999, p. 312; tr. 219.

tendency to withdraw into concealment, and thereby shelter its mystery.[30] While in *Contributions* Heidegger ponders the historical clearing of being (*Sein*) as enowning *(Ereignis*), he nevertheless casts light, relative to Da-sein, on the openness at the heart of human experience. As such, the erotic imagination illustrates this interplay between unconcealing and concealing, as the "play-space" (*Spiel-raum*) within which human beings experience their sexuality as "meaningful" as well as pleasurable. The key to eroticism resides in the element of concealment, for it is by preserving the "veil of mystery" that forms of sexual attraction, e.g., nudity, retain their allure.

Because the clearing is always overshadowed by the possibility of its opposite, the preceding passage from *Contributions* also suggests that a tendency toward closure is as much endemic to imagination, as is the impetus toward openness. Though in the *Kant and the Problem of Metaphysics*, Heidegger emphasizes the positive character of imagination, we cannot ignore the counter prospect that the power of its efficacy can also be diverted, if our phenomenological analysis is to remain true to the thing itself. In this regard, the account of the erotic imagination proves to be especially fruitful, because it illustrates how the negative tendency toward concealment is interwoven into the play-space of desire and eroticism. As is one of Heidegger's basic tenets, in every instance of unconcealment, a measure of concealment also occurs.

The factical confirmation of this "negativity" occurs in the way that the physical trigger of erotic fantasy, or fetishes, can actually usurp the creativity of the former and divert the dynamic of the "play-space" into a kind of fixation. If fetishes pose this problem, it is because, as an aid to fantasy, there is point of diminishing return as to their effectiveness. Indeed, fetishes display a distinct pathology. Specifically, their efficacy in stimulating sexual arousal requires further enhancement, *with the paradox that*, as the fantasy becomes secondary to the fetish, that is, the tangible expression of the latter, i.e., the physical icon, usurps the novelty of the former, i.e., the image–thereby undermining the source of arousal it (the fetish) would seek to sustain. If the "novelty of the possible" is the unique dynamic which ignites the erotic imagination, then its counterpoint lies in the familiarity which breeds, if not contempt, a least a kind of *ennui*. As a corollary to this ennui, the pathology of fetishes lies in fixating on the physical stimuli through this *ubiquitous presence* of the icon, to the point of spawning an addiction. The proliferation of internet pornography is a good example of how addiction arises as a result of "reducing" the fantasy to the variety of visual icons and fetishes displayed on a computer screen. Correlatively, the prevalence of fetishes, as restricting the play-space of erotic imagination, corresponds to a specific form of desire as "ontical craving."[31] By virtue of its "ontical" expression, such desire arises when the possibility of unconcealment gives way to its opposite, thereby "abandoning" the individual to the fickle attraction and allure of material things. As experienced concretely by the individual, this

[30] GA 65, pp. 389–392; tr. 271–273.

[31] For a discussion of this concept of "ontical craving," see Zimmerman 1995, pp. 501–523.

abandonment (*Verlassenheit*) to the sheer attraction of beings constitutes an "intensified" mode of the self's tendency to "fall" (*Verfallen*).

A corresponding temporalizing shapes the individual's "enthrallment" with beings (and their emerges as points of arousal or "stimuli"). The question inevitably arises as the transition from momentary arousal to habitual compulsion, e.g., from social drinker to "problem drinker" or addict. Put in temporal terms, the transition is from a "short-term" to a "long-term" fix. Going unstated, however, is that the emphasis on longevity is still a modification of the 'present,' that is, of extending its duration "indefinitely." The indefiniteness means that the present remains nondescript, blurring into a continuation of the next 'moment,' and of the next, and so on… in a fugitive search for gratification and escape. Given the illusion offered by this unowned way of temporalizing, the gate swings open for intensified falling to occur.

To be sure, Heidegger's description of the "intensified falling" of human existence casts considerable light on the problem of addiction, including that corresponding to a specific genre of sexual "fixation." In contrast to "behavioristic" models, the phenomenon of addiction cannot simply be reduced to a diagnosis of an individual's having an obsessive-compulsive disorder. On the contrary, whatever "demons" the individual may be battling, these do not rise to the level of addiction until they *overtake* the individual's power to prioritize the possibilities of his/her existence and thereby *subvert* the capacity to exercise self-concern altogether. These dual vectors of "overtaking" and "subverting" distinguish the "*worauf*" or "whereto" of the "flight" of addiction, as a mode of Dasein's intensified falling and plunge into "unownedness" (recalling the Latin root, "*addictus*," meaning to become "unowned" or "disowned").[32] Having made this link between fundamental ontology and an important topic in phenomenological psychology, let me conclude by outlining the implications of my analysis of fetishes for a phenomenological description of the erotic imagination.

The phenomenological character of our analysis reminds us that, even in the case of considering the pathology of fetishes, the "*pathos*" in question does not simply refer to a medical diagnosis, but instead to the Greek sense of the conflict, suffering, and struggle involved in each person's attempt to come to terms with his/her sexuality. Conversely, the realization that fixations occur when fetishes usurp the creativity of erotic imagination, harks back to the most basic tenet that Heidegger espouses at the outset of his phenomenology: "Higher than actuality stands possibility."[33] In term of the erotic imagination, this dictum proves to be doubly true, not only because its power originates ontologically from the self's arena of disclosedness or its "play-space," but also by "reinventing" itself by expanding the space of possibility. Within the context of eroticism, we hear the echo of Heidegger's basic tenet in the maxim that "less is more." That is, the "less" emphasis there is on the overtly graphic form of any single icon (and the more on the power of fantasy), the greater is the potential

[32] Heidegger 1977/1962, pp. 259–260; tr. 240. Also see Schalow 2006, pp. 24–26. See O'Connor 2016, p. 20 ("The word addiction means… to give over, dedicate, or surrender").

[33] Heidegger 1977/1962, pp. 51–52; tr. 63.

for eroticism and *translation* of a physical source of arousal into a kind of play. Indeed, this "*mimesis*" of our own basic physical responses, which transposes them into the wider orbit of a "play-space," constitutes the basic "spark" of eroticism beyond merely sensory stimuli. In this "mimetic" process, the image is not simply a "representation" (*Vorstellung*) of an indeterminate stimulus, but instead, as a signifying act, expands the horizon of meaning (*Sinn*) itself. The sexual fantasy, then, has its own "polymorphic" or innovative character, that "augments" or "extends" meaning, rather than just "re-presenting" it. Whether we employ Heidegger's methodology, or even, subsequently, Paul Ricoeur's,[34] the self-interpretation of how our experiences (including eroticism) become "meaningful" defines the task of hermeneutics. In this manner, the phenomenological analysis of eroticism is thereby re-enacted or "repeated" on the deeper level of the hermeneutics of being-in-the-world.

We can draw three implications of our phenomenological analysis. First, by addressing the erotic imagination, we extend Heidegger's treatment of that power beyond what can ordinarily by revealed from the point of departure of Kant's "transcendental" perspective. Secondly, by providing an alternative context to address the concern for sexuality, we outline another avenue for developing the problem of embodiment, which lies dormant within Heidegger's hermeneutic phenomenology. Thirdly, by identifying the individual's potential to become "fixated" on certain sexual stimuli, we distinguish an important corollary to the phenomenon of "addiction," the analysis of which proves to be especially germane to the field of phenomenological psychology. That is, the correlation of addiction with fetishes, suggests that the "symbolic" character of the latter extends the "hook" whereby an individual comes under the "power" of what he/she cannot control (i.e., as addicted, as "unowned")–whatever physiological basis the addictive source may have.

While this ontical focus may be annexed to the domain of "applied phenemenology," our ability to outline its importance provides a further indication of a commitment to the "*evidentiary*" character of the phenomena. Indeed, only by heeding the "concreteness" of the "thing itself," can we address a side of the imagination that is as elusive as is its potential for eroticism. In this way, the phenomenological description of imagination, and its erotic development, can be conjoined with a hermeneutic account of human existence, and thereby situated within the wider scope of the question of being, of the "to be."[35]

---

[34] See Lawler 1992, p. 69.

[35] We cannot go into further detail here into the intricacy of the relationship between human existence (Da-sein) and being. For further discussion, see Kovacs 2015 pp. 268–277.

3.4 Fetishes and Addiction

By appealing to a specific example, I have shown how fetishes can supplant the play of erotic imagination, thereby leading to the "fixation" of desire. Moreover, this dynamic may not be anything "abnormal," but, on the contrary, appears as a permutation of falling, and hence underscores the ubiquity of addiction. We have thereby shown how fetishes exhibit a unique synergy between the self's tendency to fall and the corresponding endowing of things within the environment with special significance. As our discussion thus far attests, though within our culture fetishes have become closely associated with sexual matters, the historical figure who first brought the discussion of fetishes into the mainstream was not the eminent psychologist, Sigmund Freud, but instead the philosopher, Karl Marx. And Marx proves to be significant precisely, because, as the foremost critic of capitalism, he opens up a critical counterpoint on the culture of "conspicuous consumption," which a century later defines the *Zeitgeist* of Bill Wilson's vocation as a stockbroker.

But in all his writings, does Marx say anything about addiction, whether implicitly or explicitly, or even by providing an ironic frame of reference to address this problem? First, there is the straightforward admission that the harshness of reality can become so overwhelming as to motivate people to seek "escape" from it. Indeed, the outlet may offer an immediate comfort through the euphoria of a drug, specifically, opium, or may instead hold out the *promise* of a future state of fulfillment through religious beliefs. Marx's politics aside, he makes an interesting comparison between the numbed state in which drugs put people and their receptivity to the "illusion" of believing in an afterlife. In quoting his famous line "Religion is the opium of the people," our emphasis lies not on his critique, rightly or wrongly of the former, but instead on his reference to the latter as a point of comparison. For Marx is not suggesting that addiction to opium is a pervasive problem in a clinical sense, but rather that a unique cultural climate has begun to emerge in which people have becomes so "alienated" that they have no other recourse than to seek solace in whatever can appease their suffering. Such solace may assume a "figurative" form, as the guise of religious beliefs, or something more "literal," as an actual drug.

Before "addiction" became a medical diagnosis, as it did in Bill Wilson's time, Marx's statement that "religion is the opium of the people" serves as one of the first descriptions of drugs as a form of recreational "escape." Of perhaps even greater importance, however, is the juxtaposing of an "escape" in that literal sense with an equally powerful, albeit "figurative" example thereof. For in this way, Marx suggests that escape becomes necessary not only when human beings finds themselves facing adverse physical conditions, but there is also the suffering due to "alienation," as divorced from any sense of purpose and fulfillment in life. In the four forms of alienation Marx describes, he primarily targets the variety resulting from the exploitation of the worker by the capitalist. But he also suggests that alienation occurs on the other side of equation, not only through the act of selling one's labor, but by assuming the complementary position of the consumer who disposes his/her little earnings to purchase the capitalistic "dream." In other words, the impulse to buy and consume, as an act of *indulgence* rather than of necessity, provides the mirror image of the same "icon" of fulfillment that religion does, but as a vehicle of "instant," rather than "deferred" gratification.

Marx's celebrated statement that "religion is the opium of people," when recast in terms of today's culture of conspicuous consumption, translates into an even simpler mantra: "opium is 'the opium of the people.'"[36] But opium no longer stands simply for the "drug of choice" during Marx's time (or even as today a genre of synthetic opiates), but instead signifies any and every possible source of addiction, that is, the cultural climate of addiction today. But in light of what he himself describes as a fetish, the lure and "hook" of an opiate includes another variation, which is not restricted exclusively to any substance, but instead encompasses the spectrum of process or "behavioral" addictions. Under this new rubric, sex can also become an intoxicant; when an individual becomes completely preoccupied with, and acts upon his/her sexual impulses, he/she may experience a euphoria whose overwhelming power rivals that of opium.

Insofar as Marx defined a fetish in economic terms, the flip side is that within a specific historical-cultural milieu consumption fuels addiction. Indeed, the convergence of the two becomes possible, because, as Marx anticipated, a parallel divergence of desire and needs occurs. In other words, consuming is no longer restricted to need, but instead its "boundaries" are exceeded so that consumption becomes an end in itself. As these boundaries recede, the will and opportunity for overindulgence prevails. The corresponding cultural change and transformation, as it were, then extends the hook of addiction in every direction. Where in Marx's time people may have sought solace in an illusory dream of an afterlife, today's culture of conspicuous consumption fosters a self-delusion of its own as a kind of entitlement to happiness here and now. A further transformation, and, if you were, a translation of what is signified through the hook of addiction, occurs.

When desire surpasses need, the latter does not disappear, but, on the contrary, translates into a variation thereof as addict's sense of *dependency*. To the extent that the icon "opium" translates into "the opium of the people," need becomes "dependency." That is, the addict's "need" to satisfy his/her relentless cravings overtake his/her needs in the ordinary sense, e.g., food and shelter. When the allure of a fetish translates into dependency of this kind, then *the possibility of an addiction arises–with its accompanying pathology*. What, then, is the hallmark of an addiction, its distinguishing mark? We can answer the question most simply this way. When the individual's cravings become so insatiable, the allure of the addiction fractures the individual's desires into conflicting ends, which give inordinate power to one "impulse," *only at the expense of removing the ability to control it.*

The culture of conspicuous consumption not only nurtures the cravings of the addict, but, equally as important, also creates the opportunity. In phenomenological

[36] In this connection that will reappear later on our study in impact of a further affective dimension as a "seed" of addiction, what in a psychoanalytic context defined as "trauma"—as two related, albeit separate issues. See Stolorow 2011, pp. 8–17 and Hannush's "Review" of this book, 2012, pp. 217–221.

terms, we have already pinpointed this opportunity in terms of the availability of what is "within-reach." Yet, the social criticism we have developed in light of Marx's account of the fetish of the commodity also has a phenomenological component. For the *experience* of addiction is necessarily a two-way street. The possibility of fulfilling the craving requires on its side the increase of "availability," which in turn on the other side, paradoxically, in satisfying the craving in turn fuels it. Herein lies, within our unique historical-cultural milieu, the "cycle" of addiction. The culture of conspicuous consumption extends the opportunity, in both "legal" and "illegal" directions, for access to the "opium," understood in an iconic sense, of course. In a curious way, the cycle of addiction replicates the same economic principles of "supply and demand" on which the culture of conspicuous consumption is based. The preceding observation, however, is meant purely phenomenologically. And while the language may implicate some of Marx's most vehement criticisms of capitalism, our approach does advance any ideological claims about the merits of one economic system versus another. On the contrary, the predominance of today's culture of conspicuous consumption–for better or for worse–testifies to the triumph of capitalism,[37] and, conversely, to Marx's erroneous anticipation of its imminent collapse.

Among twentieth century phenomenologists, Paul Ricouer provides one of the clearest accounts of the semiotic and iconic character of a fetish-like attachment. Employing Karl Marx's terminology, Ricoeur points to a development of "false-consciousness," which accompanies the addictive lure of the fetish. In the case of false consciousness, the individual identifies with the object of the desire, which in turn mirrors a distorted image of the self. What is desired supplants and becomes a substitute for "who" capable of experiencing, desiring and choosing, in such a way as to mask, camouflage, and conceal the potentiality of the self and source of its intentionality or capacity to engender meaning. As Ricoeur states:

> What distinguishes false-consciousness from error or falsehood…is the possibility of signifying another thing than what one believes was signified, that is, the possibility of the masked consciousness. There two words, "false consciousness," pertain usually to Marx. The metaphor of the mask is essentially Nietzschean. Consciousness, far from being transparent to itself, is at the same time what reveals and what conceals; it is this relation of conceal/reveal which calls for a specific reading, a *hermeneutics*.[38]

The mirroring of the self in the object-desired, or the fetish, falsifies "who" individual is, creating a false sense of the self in which the "external" dictates to the "internal," reversing the priority of the relation of the individual to the environment. What Marx understood as an economic enslavement produced by a false consciousness, Ricouer reinterprets hermeneutically as a captivation and ensnarement of the self to the distorted effect (and, indeed, *affectation*) of its desires. Conversely, the self-enslavement that Marx attributed to an economic system of exploitation and alienation (in nineteenth century capitalism) admits new iconic expressions, which both extends and diversifies the mode signification of the fetish (e.g., the relation between the signifier and what is signified). Put simply, the fetish-like attachment

[37] See Schalow 2000, pp. 249–267.

[38] Ricoeur 1978, p. 215.

can change into a multiplicity of new vehicles of possible addiction with a consumer driven society.

To go along with a pervasive attitude of acquiescence or "laissez-faire," the culture of conspicuous consumption devotes many of its productive energies to satisfying the "desires" of the consumer. This means that channels are set up for the distribution and "marketing," as well as production, of whatever, substance or activity, can satisfy the cravings of the would be addict. But once again the factor for "supply and demand" becomes prominent, although with the added twist of blurring the lines between legal and illegal channels of distribution. Thus, within the "celebrity" crowd, there are various channels to distribute illegal drugs, e.g., cocaine, and, albeit "illegal," nevertheless are validated within that culture and condoned despite the illegality. On the flip side, so-called "legal" narcotics, which can be distributed only through a "prescription," are nevertheless overprescribed in a way that parallels the illegal distribution of their counterparts (e.g., the overuse of "synthetic" opiates, which has led to the rise of a heroin epidemic in communities in the United States). In this way, the medical profession, including the psychiatric community, becomes the new generation of "drug-pushers." Ironically, such drug-pushing may rest on the "best intentions," because it also serves the dictates of "supply and demand," of financial profit in reciprocation for fulfilling the needs of the consumer.

We need only look at how the attempt at "medicating" away our stress has created another form of addiction through the abuse of "prescription drugs," as the tragic legacy various celebrities too numerous to mention here attests. Here is a foremost example of where the lines between legality and illegality blur in the distribution of certain drugs, indeed, so-called "legitimate" forms of treatment. We might say that the Anna Nicole Smith tragedy of death through the effects of overdose (and various other celebrities too numerous to mention) provides a noteworthy example of co-dependency, if not "enabling." But when understood in light of the culture of conspicuous consumption in general (and that of celebrities in particular), her life and death illustrates how the demand for "availability" also shifts the boundaries of the law. As a result of this shift, greater access is granted to "drugs," but at the expense of broadening the opportunity for individuals to become addicted. The link, then, between the "permissiveness" of our culture of conspicuous consumption and its laissez faire economics may be more than just coincidental. When viewed from this perspective, the addict ceases to be merely an anomaly within society, and, as immersed in the culture of immediate gratification, re-emerges as another genre of consumer. Concealed in the "wealth" of consuming is the "*impoverishment* of [our] being-in-the-world."[39] In this way, a cultural narrative about addiction becoming the "epidemic" of our age comes to light within a new interpretive horizon (beyond the rigid paradigm of contemporary psychology).

---

[39] Radloff 2007, p. 204 (emphasis mine). Also see Radloff 2014, pp. 315–330.

3.5 Conclusion

By outlining the culture of conspicuous consumption, and showing how it extends the "hook" of addiction in various directions, we have explained how (1). the fracturing of human desire occurs and (2). how the opportunities to become addicted have multiplied today. To be sure, addiction is a personal crisis, and implies individual "choices." But the problem also has a "faceless" character, insofar as its proliferation occurs insofar the self's divestiture of its individuality through addiction goes hand and hand with the investment of one's role as a consumer. By following the path of this criticism, we return to the most elemental level of the phenomenon to uncover the common origin of two distinctive forms of addiction, "substance" and "process/behavioral" arise. Having uncovered this origin, we have arrived at the double dimension of *signifying-affective* response, the birth of a fetish-like attachment (as the root of our embodiment) from which the *possibility* of addiction springs.

Given this newly cleared ground from for studying the *problem* of addiction, we have a new perspective from which to unfold the existential origins of deception as the basis on which the predisposition to become addicted *reverts into a self-destructive practice*; from this dissimulation and self-deception emerges the individual's tendency *to deny* that his/her addictive behavior constitutes a problem in its own right. In the process, we will deepen our account of the fusion of time and space as the dynamic governing both the self's capitulation to its cravings and the counter possibility of "winning back" the capacity to choose.

References

Agamben, Giorgio. 1999. *Potentialities*, ed. and trans. Daniel Heller-Roazen. Stanford: Stanford University Press, 1999.

Aho, Kevin. 2007. Gender and time: Revisiting the question of Dasein's neutrality. *Epoché* 12 (1): 137–155.

Boss, Medard. 1949. *Meaning and content of sexual perversions: A daseinanalytic approach to the psychopathology of the phenomenon of love*, trans. Liese Lewis Abell. New York: Grune & Stratton.

Cicoan, C. 2008. The question of the living body in Heidegger's analytic of dasein. *Research in Phenomenology* 38: 72–89.

Derrida, Jacques. 2001. *Geschlecht*: Sexual difference and ontological difference. In *Feminist interpretations of Martin Heidegger*, ed. Nancy J. Holland and Patricia Huntington, vol. 2001, 53–72. University Park: Penn State University Press.

Dillon, M.C. 1993. Sex, time, and love: Erotic temporality. In *Sex, love, and friendship: Studies of the society for the philosophy of sex and love 1977–1992*, ed. Alan Soble, 316–325. Amsterdam: Rodopi.

Hannush, M.J. 2012. Review of world, affectivity, and trauma: Heidegger and Post-Cartesian psychology. *Journal of Phenomenological Psychology* 42 (2): 217–221.

Harrington, Michael. 1967. *The twilight of capitalism*. New York: Simon & Schuster.

Heidegger, Martin. 1962. *Being and Time*, trans. John Macqurrie and Edward Robinson. New York: Harper & Row.
———. 1968. *What is called thinking?*, trans. J. Glenn Gray. New York: Harper & Row.
———. 1977. *Sein und Zeit*, GA 2. Vittorio Klostermann: Frankfurt am Main.
———. 1978. *Metaphysische anfansgründe der logik im ausgang von Leibniz*, GA 26. Frankfurt am Main: Vittorio Klostermann.
———. 1982. *Parmenides*, GA 54. Frankfurt am Main: Vittorio Klostermann.
———. 1983. *Die grundbegriffe der metaphysik: Welt – endlichkeit – einsamkeit, GA 29/30.* Vittorio Klostermann: Frankfurt am Main.
———. 1984. *The metaphysical foundations of logic*, trans. Michael Heim. Bloomington: Indiana University Press.
———. 1989. *Beiträge zur philosophie (Vom Ereignis)*, GA 65. Frankfurt am Main: Vittorio Klostermann.
———. 1991. *Kant und das problem der metaphysik*, GA 3. Frankfurt am Main: Vittorio Klostermann.
———. 1992. *Parmenides*, trans. A. Schuwer and R. Rojcewicz. Bloomington: Indiana University Press.
———.1995. The fundamental concepts of metaphysics. World, finitude, solitude, trans. William McNeill and Nicholas Walker. Bloomington: Indiana University Press.
———. 1997. *Kant and the problem of metaphysics*, trans. Richard Taft. Bloomington: Indiana University Press.
———. 1999. *Contributions to philosophy (From enowning)*, trans. Parvis Emad and Kenneth Maly. Bloomington: Indiana University Press.
———. 2001. *Zollikon seminars*, trans. F. Mayr and R. Askay. Evanston: Northwestern University Press.
Kant, Immanuel. 1951. *Critique of judgment*, trans J. H. Bernhard. New York: Macmillan Publishers.
———.1965. *Critique of pure reason*, trans. Norman Kemp Smith. New York: St. Martin's Press.
Kemp, Ryan. 2009. The temporal dimension of addiction. *Journal of Phenomenological Psychology* 40: 1–18.
Kovacs, George. 2015. *Thinking and be-ing in Heidegger's Beiträge zur philosophie (Vom Ereignis)*. Bucharest: Zeta Books.
Lawler, Leonard. 1992. *Imagination and chance*. Albany: SUNY Press.
Marx, Karl. 1963. In *The Marx-Engels reader*, ed. Robert C. Tucker. New York: W. W. Norton & Company.
Merleau-Ponty, Maurice. 1961. *Phenomenology of perception*, trans. Colin Smith. London: Routledge.
Mitchell, Jane Valez, and Sandra Mohr. 2011. *Addict nation: An intervention of America*. Deerfield: Health Communications, Inc.
O'Connor, Peg. 2016. *Life on the rocks. Finding meaning in addiction and recovery*. Las Vegas: Central Recovery Press.
Overgaard, Søren. 2004. Heidegger on embodiment. *Journal of the British Society for Phenomenology* 35 (2): 116–131.
Quinones, Sam. 2015. *Dreamland: The true story of America's opiate epidemic*. New York: Bloomsbury.
Radloff, Bernhard. 2007. *Heidegger and the question of national socialism: Disclosure and gestalt*. Toronto: The University of Toronto Press.
———. 2014. Contra antiphon the sophist: Aristotle, Heidegger and the planetary order. *Existentia* 24: 283–322.
Ricoeur, Paul. 1978. The critique of religion, trans. R. Bradley DeFord. In *The philosophy of Paul Ricoeur: An anthology of his work*. Boston: Beacon Press.

Schalow, Frank. 2000. Heidegger and the question of economics. *American Catholic Philosophical Quarterly* 74 (2): 249–267.

———. 2006. *The incarnality of being: The earth, animals, and the body in Heidegger's thought.* Albany: SUNY Press.

———. 2016. A Diltheyan loop? The methodological side of Heidegger's interpretation of Kant, ed. Eric Sean Nelson, 11/3. *Frontiers of Philosophy in China*, 377–394.

———. 2017. Hume and Kant on imagination: Thematic and methodological differences. In *Kant and the Scottish enlightenment*, ed. Elizabeth Robinson and Chris W. Surprenant, 213–229. London: Routledge.

Steeves, James B. 2004. *Imagining bodies: Merleau-Ponty's philosophy of imagination.* Pittsburgh: Duquesne University Press.

Stolorow, Robert. 2011. *World, affectivity, and trauma: Heidegger and Post-Cartesian psychology.* London: Routledge.

Zimmerman, Michael E. 1995. Ontical craving versus ontological desire. In *From phenomenology to thought, errancy, and desire: Essays in honor of William J. Richardson, S. J*, ed. Babette Babich. Dordrecht: Kluwer Academic Publishers.

Chapter 4
Confronting the Forces of Self-Deception

While throughout his writings Heidegger only briefly alludes to the self's potential to become addicted, he nevertheless provides several tantalizing clues on how to explicate this phenomenon in his *magnum opus*, *Being and Time*. Accordingly, the implications of my study are twofold. First, with its concrete and experiential focus, phenomenology offers an alternative way to uncover the key elements of addiction than can be found in the more conventional models of behaviorism and neuroscience. Second, Heidegger's phenomenology not only uncovers these elements, but, conversely, the phenomenological explication of them can also reinstate the problem of addiction within the larger context of the human predicament.

While there are various kinds of addiction, in simplest terms the possibility of becoming addicted begins from the fact that we are embodied. And yet, by the same token, addiction is not reducible to any set of physiological components, e.g., brain chemistry. From a phenomenological perspective, embodiment includes as much the self's openness and dynamic of projecting its possibilities, as any factical elements we might ordinarily associate with "having a body," e.g., the "handiness" by which we use items of equipment. Indeed, the ultimate thrust of Heidegger's account of being-in-the-world lies in circumventing the dualisms of mind-body, subject-object as ordinarily conceived. On the contrary, the unitary structure of being-in-the-world dictates that, while the body is an important factor, the *phenomenon* of addiction exhibits a complexity which interweaves elements of (1) our comportment toward and use of things that we encounter within the world, and have at our disposal, as it were, from our "fingertips," (2) our capacity for self-understanding (or misunderstanding, and the practices of deception implied thereby), and (3) our relations with others, which result in parasitic, "co-dependent" interactions. To be sure, all of these elements are interwoven to form a unique tapestry by which we can understand the problem of addiction. In this chapter, I will highlight the synergy between the second and third (having already addressed the first element in previous chapters); because they define the nexus in our being-in-the-world where *denial* takes hold to provide the impetus in the downward spiral of addiction. Herein lies

F. Schalow, *Toward a Phenomenology of Addiction: Embodiment, Technology, Transcendence*, Contributions To Phenomenology 93,
DOI 10.1007/978-3-319-66942-7_4

the intangible variable in the self's crossing the barrier between "social" and "recreational" use and habituated, self-destructive dependence on various substances or activities.

In the first section of this chapter, I will trace the dissimulative and deceptive practices that bring the individual to the threshold where addictive practices become habituated and self-destructive. In the second section, I will address the further confluence of environmental factors that contribute to the individual's vulnerability to addiction, including the role that co-dependency plays in shaping the gestalt of the addict's being-in-the-world. In the third section, I will outline the transmutation of spatiality into "cyberspace" as a further catalyst in extending the access ways to what is potentially addictive.

4.1 Doubling Down: Dissimulation and Deception

In seeking an alternative to the clinical approach of neuroscience, we must consider the corollary concern for the place that the (human) body occupies in our phenomenological inquiry into addiction. In other words, barring the emphasis on brain chemistry and other neurological functions, we cannot, by the same token, deny the relevance of the body in the overall "gestalt" of how addiction first arises. But then the guiding question becomes: what do we mean by the body and how does our understanding of it unfold differently within a phenomenological perspective? In this respect, we need to make the initial observation that the phenomenology places emphasis on the "lived-body," in contrast to other scientific, clinical and physiological models. But what is the lived-body and what light does Heidegger in particular shed upon this phenomenon? Ironically, throughout his writings Heidegger rarely addresses the importance of the body, and other phenomenologists have criticized him for his apparent neglect (of the "lived-body" as such).[1] Nevertheless, the inquiry into addiction may indicate an important locus of our being-in-the-world, allowing the multifaceted character of the "lived-body" can be brought to light.

For Heidegger, the body is not simply the composite of anatomical, physiological, and neurological capacities. Instead, the lived-body entails a way in which each of us is pre-oriented within the world, as a kind of niche or space of openness within the tapestry of environmental and social relations by which human beings interact on the wider stage of being-in-the-world. Accordingly, the lived-body is a dimension of our being-in-the-world and the "openness" thereof. And, conversely, the lived-body arises from, and helps us *as human beings* to navigate an expanse of possibilities, both with respect to our use of things and our interaction with others, for example, the use of hand gestures to express friendship (as in a "fist pump"). Our bodies also provide directives by which we exercise care and self-concern through the influence of and manifestations of desires, wants, and needs. Within the context of the lived-body, these are not simply the result of physiological stimuli, but instead

[1] Schalow 2006, pp. 13–23.

are interwoven into the fabric of care as harboring meaningful forms of environmental and social interaction. For example, a sexual desire is not merely reducible to a physical/hormonal impulse, but also is connected with social mores like "dating," and may even suggest an eroticism in which its development as a possibility between two people becomes paramount.[2]

The lived-body is rooted in the manner in which the self is "in" or inhabits the world, reflecting its constitutive elements of understanding, discourse, and, most significantly, moods or dispositions. One of the primary ways in which we find ourselves already situated in the world are moods and bound to a specific set of circumstances. Moods are not simply confined to subjective states or even emotions, but are a fundamental mode of disclosure by which the self becomes aware of its "to be," of its place within the world. This manner of disclosedness precedes the so-called "affective" character of moods, including how uncomfortable they make us feel. Indeed, certain moods like anxiety (*Angst*) may be intrinsically unsettling, and yet are fundamental forms of attunement (*Stimmung*), which inculcate an awareness of the fact of existing itself. As Medard Boss states: "Anxiety, being the most characteristic 'fundamental disposition of human existence' (Heidegger) is the most important barrier of the worldly-isolated mode of existence, with its limits in time and space and limitations through narrower purposes."[3] Given the unpredictability and inescapability of moods, we can speak of the self as "thrown" into the world. Due to thrownness, we are also embodied or incarnated within the flesh, and thereby thrust into circumstances that are not of our making. Stolorow makes an important observation concerning the connection between affects and the self's attuned disclosiveness of being-in-the-world. "Heidegger's term for the existential ground of affectivity (feelings and moods) is *Befindlichkeit*. Literally, the word might be translated a 'how-one-finds-oneself-ness"…*Befindlichkeit* is disclosive of our always already having been delivered over to the situatedness in which we find ourselves."[4] As already situated within the world, we are beset by a spectrum of moods and feelings, finding ourselves *to be vulnerable*, in such a way that, as embodied beings, we confront the needs, wishes, and desires preoccupying us on a daily basis. The self's pursuit of its desires, etc. may be "even-tempered" and orderly, following everyday protocol and standards of normalcy. When the desires become unruly and uncontrollable, such that the self "falls prey" to them, an ensuing chaos may erupt within the otherwise calm of everydayness. A favorable climate or atmosphere may arise within which addictions can take hold and indeed proliferate.

Having moods belong fundamentally to care, to what it means to be human. Put another way, human beings are born "moody," in the sense that they are always predisposed toward or even at the mercy of specific moods throughout their lifetimes. But moods are not simply "subjective states" of mind or mental affects. Rather, moods color our experiences as a whole, or, put another way, determining

---

[2] Kovacs 1982, pp. 211–213.

[3] Boss 1949, p. 47.

[4] Stolorow 2013, p. 453.

how whatever we encounter in the world appears to us. Moods are the self's way of being-attuned, shaping the "how" of its comportment toward things, others, and even its own sense of identity. For example, one can wake up in a "bad mood" which overshadows one's entire day. Moods reflect back upon and reveal the entirety of one's situation, and thereby disclose the entire spectrum of the self's being-in-the-world. As Heidegger states in *Zollikon Seminars*, while addressing the existential origin of stress: "Thus understood, this being burdened—the stress— belongs to the existential constitution of human being. According to *Being and Time*'s terminology (sec. 38), it is an *existentiale* and belongs to the context of the phenomenon interpreted there by the term 'falling' [*Verfallen*]."[5] In this context, we should also acknowledge how prevalent depression is both as an expression of our capacity to "fall prey" and a psychological factor in exacerbating dependency—particularly on substances like alcohol, which in itself compounds the symptoms so as to masquerade as a remedy.[6] In depression, the self withdraws into a past, which diminishes the challenge and claim of future possibilities; the malaise is both an escape into the comfort zone of the status quo, and a temporalizing in which resignation to the past undermines the promise of the future.[7]

Heidegger understood something essential about moods, about their pervasive and world-disclosive character, which has frequently been overlooked; and it is precisely this ontological aspect that speaks to the unique problem of addiction. Moods reveal the character of and state of our whole situation, speaking in a way that is often difficult to articulate in words. When anxious, an individual may not be able to state explicitly what is wrong, but nevertheless continue to feel unsettled and disturbed. "Anxiety reveals the nothing" (*Die Angst offenheit das Nichts*).[8] A mood that that is indefinite on the one hand, and outlines the silhouette of one's entire situation on the other, can also be overwhelming and fuel a tendency to "flee" or escape from one's situation. Because the unsettling character of moods and what they reveal essentially coincide, the self can easily succumb to them in a way that becomes destructive. The dynamic of "falling prey to" brings this possibility to the foreground. The impulse to escape from one's mood also brings with it a sense of dissatisfaction with one's entire situation, even though this aspect may remain beneath the surface and unspoken. And thus the tendency to engage in, or partake of something that can be mood-altering (e.g., an activity like gambling or a substance like alcohol) may become greater and greater. In this tendency to flee there is also the temptation to fall prey, to seek what is mood-altering and find solace therein. In the case of addiction, this temptation becomes irresistible and overwhelming.

Thus, the vernacular and commonsense explanation of why people over indulgence and seek solace in drugs, alcohol, etc., has an existential origin. As an apparent tautology, addicts seeks mood-altering drugs because they have moods to alter,

[5] Heidegger 2001, p. 137.

[6] For a physiological account of the "depressive" effects of alcohol on the nervous system, see Erickson 2007, pp. 121–122.

[7] Heidegger 2000, p. 142.

[8] Heidegger 1976, "Was ist Metaphysik?", p. 112.

whether to seek a "rush,"[9] a "high," a state of euphoria, or even to alleviate the kind of fundamental attunement which Heidegger describes as "boredom" (*Langenweile*). By highlighting the concern for embodiment first, and then introducing the topic of moods, we develop another perspective on the problem of addiction than can be found in neuroscience. Neuroscience locates the pleasure-centers that arouse and satisfy desires in the brain. But before pleasure can be processed, the opportunity of doing so, must be provided. The avenue for acquisition must be supplied—both figuratively and literally—in order that the individual can pursue his/her so-called internal cravings. In phenomenological terms, the opportunity to fulfill those cravings presupposes the embodiment of human beings by which we can both pursue and gain access to what we desire, that is, the proximity of whatever "those things" in the environment may be. In *Being and Time*, Heidegger describes this process of using and acquiring as "handiness," or alternatively, as accessing what is "ready-to-hand" (*Zuhandenheit*). More than metaphorically, we can refer to addiction as "using." This description reflects the fact that, environmentally speaking, the perquisite for addiction is the ability to access, make available, and "bring within reach" whatever can be the source of satisfaction and quench the desires—either as a substance or a process. The addictive impulse is also a compulsion, that is, the relentless quest to "bring within reach" whatever can satisfy the individual's cravings.

The self's narrowing of its focus of concern to what is "within reach," to whatever stands as an immediate source of gratification, creates the "space" within which addiction can occur and the additive ravings can play themselves out. As we have seen, Heidegger characterizes this way of allocating space (*Raum*) as "desevering," or the impetus to bring close or "within reach." To be a "user," then is to be adept at bringing within-reach. We cannot underestimate the importance that Heidegger's early discussion of spatiality, and the corollary issue of spatiality, has in providing an inroad to the phenomenon of addiction—as such locutions as the "ready-to-hand" or the "within-reach" attest. Phenomenologically speaking, the emphasis on handiness, the within-reach, and embodiment bring to the foreground a key aspect of addiction, or the trademark of this phenomenon, namely, the "availability" of whatever catches the fancy of the "user." From a phenomenological perspective, the capacity to bring within-reach, or "handiness," both precedes and makes possible the environmental "availability" of whatever is the source of addiction. The neighborhood "pusher" becomes possible because his/her "service" is simply the other side of the addictive impulse and desire to bring within-reach a given source of immediate gratification.

The body's capacity to bring within-reach reveals another important factor shaping our interaction with the environment. In *Zollikon Seminars*, Heidegger describes the dynamic of "bodying-forth," such that an ecstatic dimension of understanding and acting, the practical intelligence of "doing" and "performing," pervade the so-called physical employment, use, and mobility of our limbs.

[9] See Wilshire 2003, pp. 3–7.

> In each case the body always participates in the being-here[/there]. But how?....*The bodying forth [Leiben] of the body* is determined by the way of my being....Thus, bodying forth is codetermined by my being human in the sense of the ecstatic sojourn amidst the beings in the clearing [*gelichted*].[10]

In the practical intelligence and capability of its worldly comportment (*Verhalten*), the self draws upon its temporal capacity to restore, recover, and recollect by projecting future courses of action. The body's capacity to remember is not simply due to repetitive occurrences, but instead transmits the way in which our actions can be converted into habits. Just as the lived-body cannot be reduced to its physiological and anatomical functions, so memory is not just a mental operation.

The deeper, ingrained memories that are developed, and take the form of habits cannot be undone any faster than the temporal interval of their creation. The creation of memory points to a broader temporal dynamic of the self's disclosedness or openness, to its expansion and contraction. As habits develop in an overly narrow way, the self's openness also contracts, and includes only those possibilities which serve its drive toward immediate gratification. The forming of habits in a myopic way, and the corresponding narrowing of the self's field of openness, creates a climate in which 'fixations" can occur and the seeds of addiction can be planted. For example, we speak of an individual as having a "drug *habit*" or as having a "chronic drinking problem."

The contraction of the self's openness, the fetish-like craving requires a deeper level in which absconding or "fleeing" (in Heidegger's sense) of destructive behavior can take root. The habituation of addictive practices, then, corresponds to the factical, *modus operandi* whereby the self extends its tendency to flee, that is, in the evasive-turning away. The evasive-turning away of intensified falling, then, is predicated upon the self's tendency to deceive, dissemble, and dissimulate in order to perpetuate its own dependency. Examples of this self-deception include excuse-making and lying, the foremost behaviors that serves as "defense mechanisms" to perpetuate the status quo of dependency and fixation. The circle of the compulsive character of addiction closes when these deceptive practices themselves become ingrained, "second nature," and thereby "habituated." The distinctive pathology of addiction emerges in the self's fugitive enactment of its tendency to dissimulate, and thereby engage in lying in a way that proves to be pathological.

Thus, addictive pursuits rise to the level of a problem in an inverse way to the self's capacity to be forthcoming, that is, to admit the problem *as such*. In simple terms, we characterize the tenacity of the self' resistance or its recalcitrance as *denial*.

---

[10] Heidegger 2001, pp. 86–87.

4.2 Being-with Others and the Phenomenon of Co-dependency

The conventional way of asking "what is addiction?" requires to us to ask, simultaneously, "who" becomes addicted?[11] From a phenomenological perspective, our working hypothesis is that becoming addicted is connected with the individual's self-identity and the manner in which becoming addicted both begins with and leads to undermining a sense of individuality. From a philosophical standpoint, the development of various plans for treating addiction, including the first of these (e.g., the Twelve Step Program) as we will outline in a subsequent chapter, waiver by leaving ambiguous the understanding of the self, while importing various therapeutic models. But what do I mean specifically by this claim? If healing is to be part of this therapy, then the initiation thereof—barring the lack of any professional or therapeutic context—must arise from the individual him/herself, or the self broadly speaking. Indeed, the major barrier to recovery or treatment is the individual's resistance to being along this path, a penchant toward refusal or what we would in today's language call "denial." The individual must play an active role in such initiation, because only in this way is can a healing process be set in motion to counter the web of deception or dissimulating in which the tendency to become (and remain) addicted is rooted. The paradox of recovery is that *the identity and uniqueness of the individual who initiates recovery has already been masked by his/her downward spiral into addiction.*

There are may be various motivations and inducements, temptations and impulses, which lead a person to become addicted. Yet no matter what these origins may be, the downward plunge into addiction is ordinarily gradual rather than sudden. Given this observation, we are led to the simple, but nevertheless perplexing question: when does causal "using" turn into an unshakable compulsion? Or, put another way, at what point does the individual cross the line between recreational-social use and pathologically pursuing the same activity? For example, in everyday discourse we distinguish between a "social drinker" and a "problem drinker." Yet often there is a fine distinction between the two. By the same token, the ability to make this distinction already reveals something precarious about the situation of the addict: namely, how innocently an addiction to anything begins, starting as an everyday avenue of "release" and diversion, only to snowball into an uncontrollable preoccupation and indulgence.

Equally as revealing is the ease with which we discuss these different practices of becoming intoxicated, for example, "getting tipsy," or "wasted." The acceptance of such behavior as "par for the course," as the norm, points to a parallel problem: namely, that each individual already has a pre-formulated and built in defense for his/her self-indulgent behavior. Already operative is a natural defensiveness by which the individual resists any suggestion, indeed, accusation of going too far or crossing the line. We cannot overlook the role that everydayness plays in reinforcing

[11] See Levin, 1987, "Introduction," pp. 1–17.

this defensiveness on a social and cultural level. As Heidegger reiterates, the self does not emerge in a vacuum, but is already immersed in a set of social relationships as a dimension of its being-with others. By the same token, the self does not succumb to its addictions merely in isolation, but rather by yielding to a social norm of conformity or "peer pressure." Conversely, this norm provides the basis for perpetuating the addictive cycle through the "enabling" of others. Before this expression became popular, and explicitly associated with the problem of addiction, Heidegger had already described that phenomenon as an instance in which an individual "leaps in" for to falsely or presumptuously protect the other, e.g., a family member or an acquaintance from his/her self-destructive tendencies. "It [concern for the Other] can, as it were, take away 'care' from the Other… him and put itself in his position in concern: it can *leap in* for him. This kind of solicitude takes over for the Other that with which he is to concern himself."[12] For example, a spouse calls the boss, indicating that his/her spouse is too "under the weather" to make it to work today, thereby justifying the act of deceiving and lying in order to protect the livelihood of his/her significant other. Herein lies the existential basis of *co-dependency*, that is, in the presumptuous, yet the individual's self-serving "leaping in" for the others.[13]

As such, co-dependency contributes to, and provides the social climate or environment for the self's addiction to take root and ultimately flourish. In this addictive cycle, the self's entanglements in its destructive and deceptive practices becomes further reinforced and habituated. When this tendency toward self-defensiveness becomes inculcated on both a personal and social level, thereby creating a *repetitive pattern* of behavior, we speak of an individual, perhaps in the early stages of addiction, as being in "denial." The term "denial" is used by addiction specialists to describe how individuals who over indulge refuse treatment, and discount the fact that they may have a "problem" in the first place. But what is the motivation behind this denial, other than the built in self-defense to justify one's behavior and "continue using?" If we look at the problem of addiction in reverse, that is, from the standpoint of its treatment, we encounter the following anomaly: that the individual who is most in need of help turns out to be he/she who is most staunchly opposed to receiving it. On a philosophical level, this at first appears to be paradox. Yet, from a clinical perspective, this behavior, indeed, incorrigibility and recalcitrance, is "par for the course." But at the same time this tendency toward "denial" is indicative of "illness," if we want to describe addiction as such. In this respect, even addiction specialists recognize the peculiarity of addiction as a distinctly human occurrence and phenomenon.

As Drew Pinsky states: "Addiction is the only disease in which people need to be convinced that (a) they have a disease and (b) they need treatment."[14] As a result, Pinsky adds: "addiction is the predominant health issue of our time."[15] But what is

[12] Heidegger 1977b/1962 p. 163; tr. 158.

[13] See Seeburger 1993, pp. 52–55.

[14] Pinsky 2003, p. 158.

[15] Pinsky 2003, p. 159.

the origin of this recalcitrance, other than merely human stubbornness? Heidegger's phenomenology account of the "falleness" of human existence, as the beginning of the downward plunge into addiction, will shed a unique spotlight on this question. In this context, our appeal to Heidegger may be more than accidental, when we factor in his own influence on the development of existential therapy, particularly through his friend, the Swiss psychiatrist, Medard Boss.[16] In a way that points to a unusual intersection of the great minds of this period, Boss studied with Jung on the way to developing a unique approach to address the deepest conflicts of human existence. Specifically, Boss casts light on the problem of sexual obsession and fixation, indeed, on the phenomenon of "sexual addiction" before there was such a diagnosis within the medical community.[17]

Boss applies Heidegger's original concept of disclosedness in order to explain both the origin and treatment of unusual sexual fixations. According to Boss, a sexual fixation closes off the sphere of human concern, awareness, and openness, thereby creating the narrowness of focus which culminates in an obsession, e.g., on a specific sexual act or article of clothing (as in certain fetishes).[18] The source of the psychological maladjustment may be considered sexually specific. The treatment, however, must address the entirety of the individual's life-situation, the breadth of self-awareness, and the nexus of interpersonal relationships. Only by bringing back into the openness that with which an individual is narrowly preoccupied, for example, emphasizing not just the source of sexual stimuli and its object but the compass of meaningful possibilities in which one's sexuality becomes a dimension of interpersonal relationships.[19] When this therapy is successful, the individual comes to "own" his/his sexual desires and reintegrate them as a dimension of the self, i.e., of one's identity. Given this ownership, the individual can exercise responsibility toward his/her behavior and desires, and thereby circumvent the compulsive tendency to "identify" with his/her impulses.

Barring any professional or clinical intervention, the preliminary therapeutic context for the first treatment program of addiction is nothing other than the world itself, broadly speaking, everydayness, including the potential encouragement and support of the fellowship (of our addiction), that is, within the everyday context of our being-with-others. To be the initiator as well as the beneficiary of the healing process, the self appeals to its *own* "higher (or "deeper") capability that is unique to each of us but granted to all. In accord with this potential to be a self, recovery is a process of reclaiming the sense of uniqueness that is otherwise lost in succumbing to addiction. In Heidegger's terms, such (self-) recovery is predicated on *self-discovery*, in which the individual's way of "coming into his/her own" provides a new directive for existence and thereby the individual can take "ownership" for who

[16] See Heidegger 2001, pp. 138, 200–202.

[17] For a discussion of how sexual addiction can be classified as medical and psychological disorder, the statistics to show how much of the US population might be affected (e.g., 5%), see Khaleghi and Khaleghi 2011, p. 10.

[18] Boss 1949, pp. 145–146.

[19] See Scott 1990, pp. 131–141.

he/she is (including as the instigator of the healing process). Heidegger emphasizes that in this guise the self develops its potentiality beyond simply following its interests, desires, and wishes. Likewise, Jung conceives the self more broadly in terms of its search for meaning, in a way that departs from narrower and libidinous forces of power and sex that his contemporaries in the psychoanalytic school emphasized, Alfred Adler and Sigmund Freud, respectively. Yet, it was Jung's student, and confidant of Heidegger, namely, Medard Boss, who redirected the therapeutic process from the flow of everydayness, from our being-in-the-world, in such a way as to implicate a deeper philosophical understanding of the individual's potential to be a self.[20]

We speak of a person who is addicted, as if he/she is in the "grip of demons." The unique power of the addiction, such that it appear "demonic," recalls the experience from Greek lore, in which an investiture in the allure of the "thing" occurs in proportion to the individual's divestiture of his/her identity. When an individual can completely divest and "sell out" to the thing, then the latter assumes an almost "magical" power, as if casting a spell over the person. Indeed, the importance of establish the parallel between addiction and fetishes is that, in making explicit the link between the variations of "substance" and "process" of the former, we establish the "symbolic" significance whatever provides the "hook" of the specific addiction. Even in the case of a physiological dependence to a "substance," we cannot overlook the "suggestive" power and allure of what attracts, that is, as the "investiture" in the thing or its endowment with inordinate significance. As a result, the "prelude" to the act itself, the anticipation of "consuming" the substance, is also intrinsic to the "thrill" and allure of the addiction. Whatever the source of the addiction may be, for it to assume this "iconic" status indicates a comparable power of "repulsion" must arise to be break to overwhelming source of the "attraction" of the (addictive) allure.

If addiction is to assume this "iconic" status, with its superordinate appeal, then relying upon a power as fragile as the human will to break its spell is not a promising solution. Indeed, as Bill Wilson and the initial authors of the Twelve Step Program formulated this anomaly: the capacity of self-control that the addict requires most of all, is precisely what he/she is most lacking. Hence Bill Wilson, perhaps following the "hunch" of his own experiences, as well as an "inspiration," concluded the opposite of what common sense seems to suggest, namely, that the addict could not liberate him/herself from the shackles of addiction seemingly

---

[20] One of the distinct advantages of bringing Heidegger's discourse to bear on the study of addiction is that we not only illuminate a vexing problem, but also can appreciate the fragile basis on which a search for a "treatment" has historically rested. For Heidegger also stands out as a historian of thought, whose example inquiry paves the way to uncover the presumptions underlying our basic beliefs. The importing of the framework of the Oxford Group as a spiritual recipe, as we will see in Chapter Six, for a possible treatment for addiction has a "presumptive" character uniquely its own. Specifically, how can a physically rooted pathology be overcome, or at least managed, through the practice of a "spiritual" discipline? Yet, if holding to presumptions are not "wrong," it is precisely the mission of philosophy to dig underneath them and unearth both their source of truth as well as their potential for errancy.

through his/her efforts. On the contrary, the reliance upon some other "supernatural," "superhuman," power was necessary to "uplift" the addict from the depths of his/her own weakness. The "dependency" of the source of the addiction requires the individual's "reliance" upon something opposite, yet equally as "powerful." As the great philosopher, Spinoza, once remarked in referring to our "bondage" to the emotions, the only way to overcome a negative desire, e.g., lust, is by transforming it into a commensurately powerful, positive substitute, e.g., love.[21]

We can bring this part discussion to a close by revisiting" Karl Marx's famous maxim, which marks the interface between the fetish of consumerism and the escape into a illusion of ultimate fulfillment: "Religion is the opium of the people." That is, as we continue our translation of this statement, the iconic allure of addiction, in a culture of conspicuous consumption, becomes the proclamation of a new god. When opium becomes the opium of the people, the euphoria of immediate gratification provides the ultimate gateway to fulfillment. In this case, the illusion lies in the way that a material artifact (e.g., through the production and distribution of a drug) masquerades as a source of *ultimate* fulfillment and happiness. Indeed, the "fetish" character of the addiction lies in confusing the icon with the divinity, thereby converting the former into a "idol." When something of the material character becomes the focus of absolute devotion, exuding a "god-like" character, then "idolatry" reigns. As we prefigure our attempt (in Chapter Five) to distill the logic on which the formulation of the Twelve Step Program rests, its overall mission, as it were, lies in reversing the illusion of this idolatry. The false god of addiction must be replaced, substituted, and, finally, overcome as the key to a "transformed" life, which revolves around a new "center," a balanced and "humbled" self-awareness of one's "place" in the world.

4.3 The Illusion of Denial and New Avenues of Addiction

Our account of the threshold of *denial*, however, is not complete until we outline how the web of deceit and self-deception provides a catalyst for creating new avenues of addiction. In conjunction with the temporal dynamic of the phenomenon, a spatial dimension will become equally paramount, in order to exhibit the guiding premise or fore-concept (*Vorgriff*) for understanding addiction in and through the enactment of the lived-body. As we develop our understanding of addiction, we must consider more specifically the character of using as manifested in diverse ways. Not only are there different kinds of addiction, they also harbor innumerable possibilities. The traditional distinction between "substance" (e.g., drugs) and "process" (e.g., gambling) addiction gives way to new possibilities, which arise with each and every technological advance, e.g., the obsessive talking on cell phones, "texting," video games, even surfing the Internet. A new terminology emerges to describe the birth of "sex addicts" and "shopaholics." Yet, while the face of

[21] de Spinoza 1961, p. 329.

addiction continues to change, Heidegger's discussion of the *Umwelt* or environment shaping our relation to and use of things, the character of their "handiness," still remains relevant. Ironically, technological advances do not obviate this facet of our embodiment, but instead give new meaning to the "ready-to-hand" as epitomized by hand-held devices, which grant immediate access to the Internet.

The physicality of addiction involves as much the bodily extension of the within-reach (i.e., the drink or cigarette as a second appendage) as the biochemical processes of either the substance (i.e., the metabolism of alcohol or the adrenalin rush triggered by computerized day-trading). The enactment of space in this environmental way occurs within the expanse of the world, and thereby implicates the openness to which human beings belong but has already begun to narrow with the self's fallenness. On the one hand, everydayness requires this narrowing in order to accentuate the accessibility of being for our use as items of equipment. On the other hand, the "bringing close" of intensified falling becomes so predominant that the environmental field shrinks to the direct availability of any given source of gratification. Regardless of the intensity of the craving or desire, the likelihood of addiction—environmentally understood—still depends upon the availability of whatever substance or process can promise gratification. To be sure, attempts can be made to make the availability more difficult, for example through Prohibition (from 1920 to the early 1930s in the US) or the current regulation of illegal drugs. But whatever the artificial 'spatial' barriers may be, the basic *modus operandi* of the bringing within-reach (of addiction) lies precisely in circumventing these barriers and gaining access regardless of the obstructions.

Within the last two decades, recent technological advances have accentuated this desevering by providing new mediums to gain access to the same source of gratification. For example, venturing into a casino may be exhilarating to the "problem" gambler, but playing the same game of blackjack at a gambling site over the Internet may be even more captivating. Although environmental influences do not create different scenarios of addiction *per se*, we cannot discount their impact either insofar as "availability" is still a necessary prerequisite for addiction to occur. If addiction had indeed reached epidemic proportions as the "predominate health issue of our time," as talk-show host and therapist Drew Pinsky claims,[22] then cultural changes in consumer behavior, in the mentality of the "they," has fueled this recent trend. A society that welcomes stimulation on various fronts, while extolling consumption of all forms, creates more and more occasions for availability. We might then reasonably expect that such a society would spawn a greater rate of addiction among its members. A technologically based society of conspicuous consumption that fuels the impulse to buy, use, and discard, as an act of indulgence rather than necessity, offers the mirror image of the "icon" of fulfillment which orthodox religion once did, but as a vehicle of "instant" (versus delayed) gratification. Beginning fifteen years ago, the newest hand-held devices gave new meaning to the "ready-to-hand," in such a way as to place the opportunity to be addicted literally at our "fingertips."

---

[22] Pinksy 2003, p. 159.

When desevering extends into *cyberspace*, then the *boundary between availability and use begins to blur*. This transformation becomes complete when "availability" matches the individual's desire to use, and "using" becomes the primary way "to be" ahead of all other concerns. Whatever form the desire "to use" may take, whether "to be" tranquillized or exhilarated, the immediate access at one's fingertips to various sources of gratification brings the addictive phase to its most dangerous and inevitable outcome. At this critical juncture, the cycle of availability and use must be broken, save that the individual becomes completely enslaved to the craving, to the allure of his/her addiction. Even if the extent of availability can be lessened if not removed from the environment altogether, the relentless craving to bring "within-reach" and to use, the initial allure of the hook, does not recede. Indeed, the prospect of any rescue resides more deeply in the moment of discovering an alternative way "to be" that is, through the influx of new possibilities that disclose beyond the narrow pursuit of its immediate gratification. The self's opportunity "to be" *apart from* its craving to use implies an openness to existence, others, and world, which precedes the expediency of immediate gratification.

This enigma of this manner of being a "self" becomes most pronounced, for example, in the struggle to draw the line between a "social" drinker and a "problem" drinker. When interpreted in a phenomenological light, however, this "diagnostic" puzzle may provide an important clue to understanding addiction. Specifically, as great is the temptation to focus on the overtly physical and "symptomatic" side of the problem, ultimately a meaning-laden and signifying dimension creates the impetus for addiction, that is, its animating force and dynamic. For example, the hallmark of addiction may not lie only in the act of "using," but in the wholesale preoccupation with the need to use both prior to and after the actual consumption. In this sense, a "blindness" results in which the self "puts all possibilities in the service of an addiction," of "one is addicted 'towards'" [*Das "Hin-zu" des Hanges*].[23] While we normally speak of an "addiction to" something, Heidegger's emphasis on an "addiction towards" suggests to what extent there is a *directionality* and worldly character to the self's desires, rather than simply biochemical and neurological responses.

As the basic thrust of Heidegger's hermeneutic phenomenology attests, the body is not a static, physical structure. Instead, our embodiment occurs within a wider compass of concerns comprising our everydayness, and through the power of its various gestures, e.g., a wink, a handshake, shapes the encompassing structure of significance of being-in-the-world itself. As already ecstatic, the body points to the locus of the expansion and contraction of possibilities, whereby even the most basic needs and desire become meaningful or infused with significance. In providing the locus of addiction, the body yields the space of intermediation where the desires become emblematic of the self, that is, by marking directions, tangents, and diversions for its own potential fulfillment. In this regard the addictions, as diverse and multiple as they may be (e.g., from gambling to drinking, to sex to drugs, to shopping to the use of the Internet) are not merely accidental and haphazard. Instead, the

[23] Heidegger 1977b/1962, pp. 259–260; tr. 240.

addictions themselves, and the desires they emblematically exhibit, divert, and intensify, rise to the level of premium importance, thereby becoming almost "iconic." We might even say that the more disproportionately important the source of addiction becomes, the more destructive the outcome of the spiral descent will be. The case in point might be selling the services of one's body, or prostitution, to maintain a lifestyle based on whatever "substance" provides the greatest source of euphoria.

The iconic character of the "allure" entails that the self always seeks its fulfillment in whatever promises to fulfill its desires. The fact that that the desire and whatever promises to fulfill it are not directly correlated, for example, hunger and a sandwich makes the desiring in its own right primary—fueling its intensity and urgency. Addiction, however, arises not simply as the result of this ever-increasing gap between the desire and its "object," an insatiability. Rather, the tendency to become addicted stems from the concealment due to the self's falling, that is, in identifying with, yielding to, and being overtaken by its desires. This self-concealing narrows the field of one's possibilities to the pursuits which revolve around fulfilling these desires, so much so that the capacity of deferral, away from and toward alternative interests, yields to the drive toward gratification. When gratification is then dictated by its "immediacy," the dynamic of addiction emerges for the first time. To the extent that immediate gratification takes hold, and the individual becomes captive to his/her desires, the hook of addiction is set. How can we describe this "hook?"

When the concealing inherent in the they-self explicitly covers-up one's "can-be," the capacity for deferral, then falling becomes intensified. As the double concealing occurs, and the tendencies toward deception, deceit and denial prevail, the spiraling down of addiction begins. Without this capacity for deferral, or what in the vernacular might be called "delayed gratification," the desire becomes indistinguishable from its object—and fixation sets in. The fixation on the objects reveals the degree to which the desire has reached the apex of its iconic status. In the process, the urgency and intensity of desiring conceals itself to the point that now only what appears is the source of gratification itself. When the object itself becomes captivating in its own right, and thereby assumes a power that can monopolize an individual's concerns over and above everything else, we can say a "fetish-like" attachment occurs. This fetish-like attachment, in which the individual wishes to "possess" takes possession of him or her, constitutes the hook of addiction. In phenomenological terms, the fetish-like attraction exaggerates and energizes an individual's desires in a specific direction. The desiring becomes a craving, in which a synergy occurs between the self's tendency to fall and the correlative endowing of things in the environment with added significance. In a fetish-like manner, the focus of the fixation both stimulates and overpowers, as if casting a magical spell upon or incantation over the addict, and yet, correlatively induces an overpowering desire to consume that in itself consumes and overpowers the individual (prior, during, and after) the course of using itself. When desire surpasses need, the latter does not disappear, but, on the contrary, translates into a variation thereof as the addict's sense of *dependency*. That is, the addict's "need" to satisfy his/her relentless cravings overtakes his/her needs in the ordinary sense, e.g., food and shelter. When the allure of a fetish translates into a dependency of this kind, the possibility of an addiction arises—with its accompanying pathology.

What, then, is the hallmark of addiction, its distinguishing mark? We can answer this question most simply this way. When the individual's cravings become so insatiable, the allure of addiction becomes so strong as to fracture one's selfhood into conflicting ends [(e.g., good health versus immediate gratification], thereby giving inordinate power to a single impulse and *at the expense of removing the ability to control it.* As the individual's capacity for disclosure centers exclusively on the object, the expanse of openness narrows to uncover only those possibilities, which further the ends of addiction and provide access to the sources of immediate gratification. The restriction of the self's openness as an "ontical craving of beings" entails a complete submission and attachment to these sources of gratification and arousal.[24] The fetish-like captivation by and attachment fuels the incessant craving of the addict, and, simultaneously, magnifies the destructive tendencies to lie, deceive, and deny in order to sustain the addiction. Because falling, and the concomitant behaviors of lying, deceiving, and excuse-making belong uniquely to human existence as care, human beings can descend into the lowest levels of depravity. In falling, one forsakes one's identity, as unowned or inauthentic, for an illusory substitute, e.g., the histrionics of gambling, which mirrors a desire to recapture the unity of the self that is otherwise lost due to the fragmentation incurred from being lost in the "they." But such attempts to recapture one's identity only amplify the self's fleeing, fugitive tendencies, which fuel the craving(s) of addictions.

We must make explicit the key precept in Heidegger's description of our openness (and restriction thereof) in order to complete our account of the hook of addiction and its distinctive dynamic: namely, the role of temporality. In the anticipatory movement of unfolding the future, and coming back to rediscover the past, temporality makes possible the ecstatic openness of the self (in and through the fullness of the moment). Because the dimension of the future takes priority, in conjunction with the past, this temporal dynamic allows the self to pursue possibilities with long-term vistas, rather than actualize those on the basis of impulse and expediency. Given these vistas or horizons, the self can cultivate possibilities, both long and short term, of different level of importance and thereby set priorities for the development of its own existence. These priorities, and the temporal dynamic on which they are predicated, empower the "can" of deferral through which something like "delayed gratification" becomes possible.

But the converse also holds in terms of temporality. If the circular dynamic of temporality comes to fruition in the owned, authentic self, then the falling of everyday, linear time equally makes possible inauthentic, unowned selfhood. Unlike original or primordial time, the derivative, falling way of temporalizing centers on the 'present,' of making present whatever appears in the proximity of one's concern. In the intensified falling of addiction, the 'present' becomes even more prevalent by providing the occasion to fuse the craving and its source of gratification. As if by definition, the immediacy of gratification necessarily revolves around the 'present.'

[24] Zimmerman 1995, p. 503.

Moreover, the iconic, fetish-like character of the addiction dictates an expectation to make present, the gratification itself, even while the actual occasion of "use" may not be imminent, that is, as a pre-occupation which still "awaits" the present as an occasion to use again. By implication, we can also say that the future becomes parasitic upon the present, as merely providing the next occasion—and so indefinitely—to use. By the same token, the past establishes the pattern of use, the inevitability of repeating the same act, and in a way harboring the iconic character of the desire, the nostalgia for using once again. Without the fugitive, escapist character of the past, the obsessive-compulsiveness of addiction would be lacking. Indeed, addiction occurs as a faulty reconstruction of the routine of everyday lie, in which its regiment is not defined by the entire course of one's day but instead by a single activity—drinking or gambling. Herein lies the existential basis for the habituality of addiction.

In unownedness also resides the possibility of ownedness. For only by confronting the downward spiral of falling can the self also embrace the challenge of seeking to be an individual and claim "ownership" for its existence, including the capacity to choose and act. According to Heidegger, authentic or owned existence occurs when the self temporalizes through its anticipation of the future by cultivating those possibilities that disclose its uniqueness and exhibit its finitude. Conversely, addiction closes off the self's future, restricting it to the next opportunity to pursue a source of immediate gratification, in short, to "get a fix." By closing off the future, the fixity of addiction constricts the original openness of human existence. Because concealment and denial fuel addiction, the addict can overcome those tendencies only by cultivating this openness, which, however, remains foreign to him/her. Put another way, the predilection to be responsible, to be answerable, is what the addict possesses least, but requires most.

From an ontological perspective, the possibility of recovering from addiction appears to harbor something of a paradox. On a practical level, breaking the cycle of addiction is equally problematic, insofar as the prospect of a "relapse" is more the norm than the exception. The practical and the philosophical points of view converge, however, insofar as we ask: where does the pathway to recovery arise, if not through a temporal indicator which points back to the origin of addiction in and through the temporalization of falling? Specifically, recovery is an ongoing task that "each day" summons the individual to confront the "can be" of existence and his/her potential to choose him/herself again, i.e., in sobriety. AA's chief mantra of recovery, "one-day-at-a-time" proves to be efficacious as a guideline for treatment,[25] precisely by approaching a life-crisis as it originates from its deepest root in human temporality—the self's unownedness of seeking immediate gratification in the present by escaping into an illusory future and fugitive past. The strategy of recovery, then, presupposes a dynamical view of the self. According to this view, the self exercises its freedom by acknowledging precisely those limitations which are set by one's temporal finitude: that one's power to refrain from addictive behavior, e.g.,

[25] B. (Dick) 1998, p. 136.

drinking, does not extend beyond the "day" and hence must be appropriated anew, i.e., *made one's own, each day forward.*

Whatever the methodology of treatment may be, it always begins with the self's own initiative to be liberated through its openness, the open resolve (*Entschlossenheit*) of accepting the abundance of each day, including the restorative, "curative" (in the sense of care) power of one's own body. For the "release" from the cravings and fixations of the body is another permutation of Heidegger's concept of authenticity or ownedness, which, as Guignon emphasizes, is thought from the perspective of the lived-body and its "being part of nature" as an "expanded conception of Dasein's finitude."[26] To be authentic is also to take "ownership" of one's body, as a further dimension of "coming into own," including the challenging task of confronting one's deeply rooted feelings, managing them as a stage of self-discovery, and even cultivating good health (e.g., through nutrition and exercise).[27] Such simple ways of exercising care require the self to set new *priorities*, in order to redirect its existence along a more fruitful and constructive path.

More than simply a bad habit, addiction implicates the entire care-structure of human existence. By recalling this primordial fact, we avoid any simplistic models of addiction that turn the phenomenon into either a fatalism of chemical processes of the brain or a voluntarism of the will's failure to exert mind over matter. The phenomenon of addiction cannot be reduced to different forms of privation: either to the physical plane of something amiss with the addict's brain chemistry or the spiritual-moral plane of something lacking in the addict's exercise of self-restraint. On the contrary, as an expression of human *pathos*, of its vulnerability to trauma and suffering, addiction implies a "genuine existential situation and spiritual distress";[28] such a multi-dimensional crisis cannot be resolved without a richer disclosure of meaning and development of a new life-trajectory. Thus, wherever the "fault" supposedly lies—and the corresponding treatment thereof—the addict must still address the question of the meaning of life and wrestle with his/her plight against the existential backdrop that he/she shares in common with the non-addict. Insofar as the potential to be addicted unfolds on this existential stage, our temporal-spatial analysis of the dynamics of addiction reveals how intricate, complex, and encompassing this phenomenon is.

Our phenomenological approach to studying addiction, and its development into a contemporary crisis, centers on the wider spectrum of the human condition, rather than what can be attributed psychologically as a proximal cause, e.g., trauma.[29] Nevertheless, as it is *experienced*, the violation of the *boundaries* that safeguard

---

[26] Guignon 2000, 96.

[27] For further discussion, See James Aho and Kevin Aho 2008, pp. 151ff. pp. 149–154.

[28] Kovacs 2013, p. 315. On a positive side, trauma can provide a catalyst for disclosing the self's finitude, despite, or even because of, the negativity of its affect and the accompanying attunement of anxiety (*Angst*). See Stolorow 2013, pp. 454–455.

[29] Carolton Erickson suggests that for women trauma may play a greater role as a causative factor in chemical dependence, thereby raising the question of "gender-specific" treatment. Erickson 2007, p. 212.

one's individuality, whether at an early stage of childhood or subsequently, for example, as induced under the guise of "PTSD" (post-traumatic stress disorder) in war veterans, interrupts or suspends the deeper capacity to exercise care. As a result, some therapists have advocated a method of "dual diagnosis," whereby emotional, social, and spiritual crises have to be included to addressing the root of addiction, e.g., a diagnosis of clinical depression as accompanying a specific addictive disorder, e.g., alcoholism.[30] Any such experience that tears apart the fabric of care, with all its accompanying symptoms of depression, forlornness, melancholia, and insomnia, still implies uneasiness, the dis-ease, of facing the ambivalence of our finitude.[31] As Kemp points out, melancholia and depression may pervade the addict's comportment toward the basic concerns and necessities of life, including avoidance in carrying out tasks like paying one's bills in a timely manner.[32] These dispositions may become so pervasive as to disable the individual's power to make simple choices in conducting his/her business in the world. In his lecture course from 1929/1930, Heidegger first describes the fundamental attunement, whereby the self holds itself within the tension of this ambivalence, "homesickness." For Heidegger, homesickness is the heart-felt struggle waged at the boundaries of what it means to be human, and thereby provides the wider horizon against which the individual can experience specific emotional distress, e.g., trauma. If trauma can be identified as a root-cause of addiction, as a tendency to "take flight" and escape,[33] then homesickness transposes the self into the center of a struggle between hope and despair, recovery and resignation. The individual who slips into the depths of denial can also rise toward recovery by embracing his/her own finitude and the openness of self-transcendence.[34]

4.4 Conclusion

By addressing the self's tendency toward deception and dissimulation, we have now outlined the process of intensified falling, that is, by which the individual crosses the threshold of denial and enters a new "zone" in its vulnerability to environmental stimuli, the "fetish-like" allure from which addiction arises. What happens when this "zone" is permeated by the intrusion of cyber-space? Can technology accelerate the movement of falling, creating a *vortex for the plunge into addiction* by mediating human desire on a global scale? Let us consider how the development of digital

---

[30] Khaleghi and Khaleghi 2011, pp. 18–20.

[31] Romanyshyn and Whalen, in Levin (ed.), 1987, pp. 189–220. This article provides an excellent account of the "American dream" as forming the backdrop of Bill Wilson's culture, the tension between the struggle for material success and the "uprootedness" of urban life.

[32] Kemp 2009, pp. 1–18.

[33] See Pinsky 2003, pp. 197–198.

[34] Heidegger 1985. In his lectures from the Summer Semester of 1936, Heidegger describes the self as perched on the precipice of the "highest heaven and the deepest abyss" (pp. 177–178). Also see Heidegger, 1977a, pp. 31–34.

technology becomes a new vehicle for blurring the boundaries of everyday spatiality, thereby extending the channels of availability for multiple sources of fetish-like attraction.

References

Aho, James, and Kevin Aho. 2008. *Body matters: A phenomenology of sickness, disease, and illness*. Lanham: Lexington Books.

Boss, Medard. 1949. *Meaning and content of sexual perversions: A daseinanalytic approach to the psychopathology of the phenomenon of love*, trans. Liese Lewis Abell. New York: Grune & Stratton.

de Spinoza, Benedict. 1961. Ethics. In *The rationalists*. New York: Random House.

Dick, B. 1998. *The oxford group & alcoholics anonymous*. Kihei: Paradise Research Publications.

Erickson, Carlton. 2007. *The science of addiction: From neurobiology to treatment*. New York: W. W. Norton.

Guignon, Charles. 2000. Philosophy and authenticity: Heidegger's search for a ground for philosophizing. In *Heidegger, authenticity, and modernity: Essays in honor of Hubert L. Dreyfus*, vol. 1, ed. Mark Wrathall and Jeff Malpas, 79–100. Cambridge: The MIT Press.

Heidegger, Martin. 1962. *Being and time*, trans. John Macquarrie and Edward Robinson. New York: Harper & Row.

———. 1976. Was ist metaphysik? In *Wegmarken*, GA 9. Frankfurt am Main: Vittorio Klostermann.

———. 1977a. *The question concerning technology and other essays*, trans. William Lovitt. New York: Harper & Row.

———. 1977b. *Sein und Zeit*, GA 2. Frankfurt am Main: Vittorio Klostermann.

———. 1985. *Schelling's treatise on the essence of human freedom*, trans. Joan Stambaugh. Athens: Ohio University Press.

———. 2000. *Elucidations of Hölderlin's poetry*, trans. Keith Hoeller. Amherst: Humanity Press.

———. 2001. *Zollikon seminars*, trans. F. Mayr and R. Askay. Evanston: Northwestern University Press.

Kemp, Ryan. 2009. The temporal dimension of addiction. *Journal of Phenomenological Psychology* 40 (1): 1–18.

Khaleghi, Morteza and Karen Khaleghi. 2011. *The anatomy of addiction: Overcoming the trigger that stands in the way of recovery*. St. Martin's Press.

Kovacs, George. 1982–1983. The personalistic understanding of the body and sexuality in Merleau-Pony. *Review of Existential Psychology and Psychiatry*, 18: 207-217.

———. 2013. Skepticism and *logos* in Nietzsche and Frankl. *Existentia* 23 (3–4): 309–324.

Pinsky, Drew. 2003. *Cracked. Putting broken lives back together*. New York: Harper-Collins.

Romanyshyn, Robert B. and Brian J. Whalen, 1987. Depression and the American dream: The struggle with home. In *Pathologies of the modern self*, ed. David Michael Levin. New York: New York University Press.

Schalow, Frank. 2006. *The incarnality of being: The earth, animals, and the body in Heidegger's thought*. Albany: SUNY Press.

Scott, Charles E. 1990. Heidegger and psychoanalysis. *Heidegger Studies* 6: 131–141.

Seeburger, Francis. 1993. *Addiction and responsibility*. New York: Crossroads.

Stolorow, Robert D. 2013. Heidegger and post-Cartesian psychoanalysis. In *The Bloomsbury Companion to Heidegger*, ed. F. Raffoul and E. Nelson, 451–458. Bloomsbury: London.

Wilshire, Bruce. 2003. *Wild hunger: The primal roots of modern addiction*. Lanham: Rowman and Littlefield.

Zimmerman, Michael E. 1995. Ontical craving versus ontological desire. In *From phenomenology to thought, errancy, and desire: Essays in honor of William J. Richardson, S. J.*, ed. Babette Babich. Dordrecht: Kluwer Academic Publishers.

Chapter 5
Technology and the Rise of the Artifice

Today, technology has emerged so prominently on a global scale, as to become almost inconspicuous in its use. Through the use of mobile devices, most notably, the iPhone, technology appears almost as an extension of each of us. Not only does technology open up new avenues by which individuals can become addicted, a person's over reliance upon such hand-held devices can itself create a unique *dependency* in its own right, that is, become "addictive," if only in a metaphorical sense. Even briefly interrupting the use of "smart-phones" may yield a host of phobias and anxiety-disorders. In simplest terms, the development of the Internet (in the 1990s) and its portability through the advancement of "cellular" technology a decade later, has proven to be a game-changer. The access to the "information superhighway" in the palm of one's hand opens up a new Pandora's Box of entertainment and diversions, which dwarfs the impact of television (if only by providing digital access to it through one of many "applications").

Even those intellectuals who possessed the greatest foresight in the twentieth century—from Huxley to Heidegger—could not have foreseen how ubiquitous the *everyday* use of technology would become from second-to-second. When Bill Wilson began to formulate the first plan to treat addiction, technology, as we know it today, was in its mere infancy. To put things in context, all we need to recall is that the invention of the television was almost two decades away! What we assume today as commonplace, from the Internet to the personal computer, were all but a dream in Bill Wilson's day. In the late 1990s, the "day-trading" of securities over the Internet became the newest craze (indeed, with addictive implications), when, some seventy years previously, during the great stock-market crash of 1929, New York Stock Exchange transactions were reported by a crude "ticker-tape." But the rapid advance of technology, and its proliferation into the everyday household, brings danger as well as convenience and efficiency. For Bill Wilson's generation would be the first to witness the destructive potential of atomic weapons, and live in the shadow of the possibility of a nuclear holocaust. The technology of today, however, may be measured more in terms of its close alignment to the culture of conspicuous

F. Schalow, *Toward a Phenomenology of Addiction: Embodiment, Technology, Transcendence*, Contributions To Phenomenology 93,
DOI 10.1007/978-3-319-66942-7_5

consumption. More than ever, Madison Avenue sings the praises of the latest technological "gadgets," from "smart phones" and microwave ovens to Ipads and high-definition televisions.

Yet, whether any of today's social problems can be attributed to the exponential growth of technology, today's culture is more readily inclined toward addictive disorders than ever before. For example, "shopping addiction" constitutes one of the newest of these disorders. But would such a problem have even been possible without the increase in the availability, not just of products, but purchasing power? Without the technological conveniences of credit cards, the buying and selling of products over the Internet and television – including the growth of such outlets as the "Home Shopping Network"—it is difficult to envision that such a pervasive problem as shopping addiction would ever have arisen. Indeed, the consumer mentality is aided and abetted by the drive to bring "within-reach," to immediate proximity, whatever attracts, allures, captivates, and eventually "hooks"–that is, the addictive substance or activity. Where we earlier alluded to the "fetish of the commodity" to describe this allure and "hook," we discover the development of a converse scenario: that is, the "*commodification* of the fetish," which technology unleashes within the arena of capitalism.

In this chapter, I will examine, first, the acceleration of the addiction "crisis," as it broadens its reach within the technological culture of the Internet, and its global distribution of the opportunity or "means" for more people to become addicted. By expanding Heidegger's account of the hidden *danger* of technology, and particularly showing the further influence of cybernetics upon our culture through what Debord calls the "society of the spectacle," I will outline the rapid cultural changes that have commercialized our fetish-like urges and developed a global market-place for their pursuit. Ironically, the more economically secure many of us become in the basic necessities of life, the more easily we may also be victimized by our self-indulgent desires and diversions, e.g., from gambling to the use of "recreational drugs." We need look no further than the "celebrity culture" and the rise of addiction among the more affluent as well as the poorer classes of society. Today's cultural milieu of addiction, and the "enabling" of addictive practices throughout various facets of society, has become pervasive. By the same token, it becomes increasingly difficult to gain a critical perspective on this problem, because our own diversions (and even our life-styles) are inextricably bound to those same technological advances that have brought all kinds of potential sources of addiction "virtually" to our fingertips. How does technology appear in this guise? *Put simply, technology appears as the inconspicuous backdrop for the mediation of human desires on a global scale, assimilating them into a vortex of self-indulgence and immediate gratification.*[1]

In the second part of this chapter, I will show how this *gestalt* of technology spawns new kinds of addictive disorders, e.g., video games, which preys upon a cultural malaise of boredom, on the hand, and, on the other, a penchant for fleeting

[1] Don Idhe employs the description "quickly mediated distance" to describe the phenomenon of cyberspace. See Ihde 2010, p. 137.

instants of excitement. We will thereby offer further testimony as to why addiction has become (in Pinsky's words) the new "epidemic" of today. Rather than assume its importance, *we must translate that narrative into philosophical terms, and thereby allow the cultural phenomenon of addiction to speak (i.e., to "show itself") as a referendum about the danger which technology poses to us on a daily basis*. In this regard, addiction ceases to be merely a clinical malady, and instead becomes a *signpost to exposing a pervasive, albeit hidden danger posed by the assimilation of our culture within a technological framework* (*Ge-stell*). Finally, in the third part of this chapter, I will illustrate how today's treatment (and "rehab") programs come directly into conflict with the culture of excess and conspicuous consumption, whose priorities for "immediate," quick solutions are precisely the opposite of a "tempered," "step by step," slowed pace of healing and recovery.

5.1 The Shadow of Technology

5.1.1 The Development of Cybernetics

Historians of philosophy credit Martin Heidegger for spearheading many key intellectual movements in the twenty-century, from developing a terminology that would be largely adopted by existentialism, to resurrecting the perennial, albeit perplexing question of (the meaning of) being. But one overarching insight that we cannot underestimate is Heidegger's acute glimpse into the troubling onslaught of technology. Indeed, if decades from now someone were to write an intellectual history of the twentieth century, a chapter would have to be reserved to the prophetic vision that Heidegger demonstrated as one of the first thinkers to address the danger of our technological age, which he initially described by the term "machination" (*Machenschaft*).[2] By "machination," he did not simply mean the use of machines and gadgets, or even the technological innovations that led to their development. Rather, Heidegger has in view a complete change in humanity's relationship to everything that is, including itself and nature. As a result of this transformation, which occurs on a global scale, humanity and nature become intertwined as "resources" to be used and exploited for exclusively instrumental purposes.[3] Thus technology is in itself a complex phenomenon, which becomes increased nuanced through its transformation into cybernetics (beyond the Industrial Revolution of the late nineteenth and early twentieth century through tool-use and manufacturing).[4]

We need only think of the current "oil" crisis to conjure up an example of technology's drive toward planetary exploitation and domination. As the "resource" of all resources, the competition for oil is not confined to the need for petroleum

---

[2] Heidegger 1989/1999; pp. 128–140; tr. 90–98.

[3] See Heidegger 1977a, pp. 1–35.

[4] Heidegger 2003, p. 63.

products, but rather its global necessity drives companies, and ultimately nations, into ruthless competition and conflict with each other. This struggle may even precipitate wars. By the same token, the individual becomes assimilated into this global network of instrumentality, the proverbial cog into the machine, divesting his/her identity on the one hand, whose value becomes reduced to his/her labor as a "resource" to be exploited. As Heidegger remarks in his pivotal essay, "The Question Concerning Technology":

> .... A tract of land is challenged into the putting out of coal and ore. The earth now reveals itself as a coal mineral district, the soil as a mineral deposit....Agriculture is now the mechanized food industry. Air is now set upon to yield nitrogen, the earth to yield ore, ore to yield uranium, for example, uranium is set upon to yield atomic energy, which can be released either for destruction for peaceful use.[5]

At the heart of Heidegger's thesis is his claim that technology "reveals" as it "conceals." Humanity's inability to question the latter, in seeking blind acceptance of the former, constitutes the "danger" of technology. As Heidegger states in his essay, "The Danger" (1949):

> The essence of technology, enframing, conducts its own disguising.
>
> One is also relinquished to this self-disguising of enframing when, at time, one darkly gleans and clearly admits for a moment that... technology itself would now draw the human after it as its instrument, be it that the human blindly follows the wrestling away, be it that he unflaggingly strives to put technology to healing and beneficial purposes.[6]

Concealed in all that is revealed in the benefits of nineteenth and twentieth century industrialization were the side effects of air pollution. Through machination, technology reveals (all that is) within the one-dimensional compass of instrumentality; but this narrowness of this revealing is predicated on a commensurate degree of concealing. Thus, the forest may be revealed for its instrumental value as potential lumber for the mass manufacturing of suburban homes. But simultaneously the forest (as belonging to nature) is concealed in its environmental significance, in the majesty of its beauty not to mention its symbiotic relationship in purifying the air for animal life (including human beings). In the digital age, technology unleashes new possibilities of instrumentality, particularly as connected to the storing, processing, and distribution of "information." *The so-called "information super highway" defines the rise of the artifice.*

The rise of the artifice is itself a primary example of a "revealing" that simultaneously conceals, and spawns the "untruth" of equating "what is" to a media *constructed* play of images.[7] Following the lead of the American scientist, Norbert Wiener, Heidegger had referred to this development as cybernetics, although he never lived to see the creation of the Internet.[8] For example, while sexuality may be

[5] Heidegger 1977a, pp. 14–15.

[6] Heidegger 2012, pp. 57–58.

[7] The artifice is the permutation of the "*techne*" of technology, which makes explicit the one-dimensional character of production that extends into cyberspace to include the manufacturing and distributing of images.

[8] See Polt 2015, pp. 180–182.

considered a "natural" act, and hence a meaningful expression of our embodiment, the proliferation of sexual images over the Internet (and even through the medium of advertising) constitutes a digitized form of "production," i.e., a form of the artifice. The cyber image becomes an iconic representation of what we may ordinarily experience as a source of stimulation in our environment, thereby mediating our desires through a graphic play of imagery (whether on a computer screen, phone or "tablet"). Of course, the early form of television was the precursor to such cyber technology, albeit lacking the ubiquity that the digital connection to the Internet provides. The ubiquity of the Internet, then, tailors our impulses to the newest and most graphic imagery (of stimulation), thereby hyperextending the locus of both desire and gratification. The "net" or web of the Internet spirals outward to captivate our desires in increasingly specialized ways, indeed, through a play of imagery that both preys upon and extends our fetish-like urges. This spiraling of the Internet web through its various algorithmic applications—the hallmark of cyber technology—provides a stage for revealing the spectrum of our desires. But does such revealing also include a counter pole of concealing?

The mode of revealing that occurs through a global network for the instantaneous exchange of information conceals the *danger* of constricting the field of human interest to the basest forms of self-indulgence or immediate gratification, of which the artifice of sexuality and eroticism is a foremost example. With the rise of the artifice, the individual becomes more vulnerable to the forces of use and exploitation then those that first appeared in the "machine-age." In *Contributions to Philosophy*, Heidegger equates machination (*Machenschaft*) with the development of a global network of human interaction and the mobilizing of diverse resources, including labor. In the grip of machination, modern culture displays three basic dynamics: (1). "massiveness," "calculation," and "acceleration" (*Beschleunigung*)[9] In this regard, "massiveness" distinguishes the overwhelming manner in which human beings experience the power of machination, which belittles everyday activities and skills (e.g., works of craftsmanship) in favor of the proximity and convenience of various gadgets.[10] Most troubling is how "the gigantic" (*das Riesenhafte*) compresses the focus of human concern by *positing* as important for the "subject" that which by another, incalculable measure (*Maß*) is indescribably trivial, thereby providing a platform to exercise the most narcissistic urges. In calculation lie the seeds of the Promethean drive of cybernetics, including a digital matrix for imposing control through the instantaneous gathering and processing of information. In this regard, *the overwhelming reach of the gigantic provides the cornerstone for the rise of the artifice*, for example, through the transmission of pre-packaged information via *hand-held devices*, including the gestalt of social media. What can be more narcissistic than the impulse to "text" or read emails, etc., while driving an

[9] Heidegger 1989/1999, pp. 120–124/84–86.

[10] Heidegger 1989/1999, pp. 136–139; tr. 95–96. For further discussion of machination, and the "claim of massiveness," "calculation," and "acceleration," see Emad 2013, pp. 43–56. As Emad emphasizes, for Heidegger machination arises in accord with the nihilistic trend of the abandonment of and by being (*Seinsverlassenheit*).

automobile—an activity now common place even if condemned by our political leaders? Finally, the development of cyber technology spurs the third dynamic of machination, namely, acceleration, that is, the impetus for the frenetic pace in which we live our lives and attempt to cram everything we can—down to sending the last email—into a twenty- four-hour day. Clock-time, which whittles the 'now' down to a microsecond, predominates as the primary mode of temporalizing,[11] exacerbating the individual's tendency to become vulnerable to its desire for immediate gratification.

Whether as a worker on the oilrigs, an infantryman serving in a war, or the driver "texting" in traffic, the individual's identity begins to dissolve in the various ways in which his/her attention are be diverted, much less his/her abilities can be used and exploited. To be sure, there are economic forces at work in these different examples of exploitation and diversion. Yet, Heidegger's prevailing insight is that global reach of technology both creates the possibility for, and fosters a climate within which to unleash these forces. In other words, the growth and expansion of technology gives a structure to the competitive practices to capitalism. Put another, by extending the free market system of buying and selling across the four corners of the world, technology creates the conditions in which capitalism can emerge as a global system, rather than as merely a national ideology. When capitalism achieves the global outreach, in which anything that can be consumed for better or worse, can also be produced, distributed and sold, including (and most notably) the assorted "trappings" of addiction.

Heidegger coins the terms "enframing" to describe the overall matrix of ordering and organization—allowing the "gigantic" to expand on a global scale—which reduces everything to its instrumental use-value. Within enframing, human beings not only use technological gadgets; they do so at the risk (e.g., the potential "danger") of all their pursuits and activities becoming assimilated into a plan which grants priority to the (instrumental) ends of expediency and immediate gratification. The users become vulnerable to exploitation in the marketplace both as producers and consumers. As Richard Polt emphasizes: "Instead of meditating on the gift of being, we run the risk of becoming entangled in the technological mode of disclosure, so that we find no possibilities for ourselves other than willing, representing, and manipulating—and no possibilities for things other than to serve as fuel and raw material for our activity."[12] Yet, more telling, enframing sets forth and posits all human activities within a single *time-frame*, that is, the instant, compressing all that we do into what can accomplished, done, enjoyed as quickly as possible. With the emergence of digital technology, the desktop computer let alone various hand-held devices, the boundaries between work and entertainment blur. The compression of human interests, of our concernful activities into the moment of immediate gratification, gives license to the allure of the self's pursuit of its fetish-like attachments. Enframing enframes the stage for tool-use,[13] granting *ubiquity* to its contemporary

[11] See Aho 2009, pp. 116–118,

[12] Polt 2015. p. 185.

[13] Zimmerman 1990, pp. 221–227.

application through digital technology and thereby sanctifying the allure of whatever serves the ends of immediate gratification, including the fetish-like attachment of addiction. Under the auspices of enframing, digital technology assimilates the culture of conspicuous consumption, thereby creating a global platform on which the seed of addiction can be planted.

With the exponential advance of digital technology, the cycle of production and consumption—and all of its conduits—can unfold on a global scale. As we saw at the close of Chapter One, Bill Wilson first confronted his own demons of addiction on this cusp of a new culture of conspicuous consumption during the "roaring 20s." But the capitalism of the "roaring 20s" was a far cry from the global force that it would ultimately become in our computerized, digital age. Indeed, if Bill Wilson's profession as a stockbroker stands as a symbol of a rampant materialism, then the culture of immediate gratification is only one step away. For all that is required to complete this next step in the march of history is the development of mouse pad and a cell phone, which places the satisfaction of all our needs and desires, literally, at the edge of our fingertips. Now the logistics of "distribution" radically shift to what can be "processed" through a "transaction" over the Internet, whether bidding on products on QVC or the Ebay, gambling on video poker or football games, or downloading pornographic film clips. A phenomenon arises, whereby not only process or "behavioral" addictions become prevalent, but the circumstances are created for the intersecting of various forms of addiction. Where smoking and drinking once went "hand in hand," now we see the arises of "multiple" addictions, compulsive buying, pill-popping, and promiscuity.

And what about the spatial dimension of the "within-reach," of the various addictions themselves; does a significant change occur with the creation of "cyberspace"? As the channels of distributions change, so does the idea of "availability," on which much of the "culture" of addiction is predicated, change as well. The "within-reach" extends beyond the locus of physical proximity, to include instead the "processing" of transactions over the Internet. In certain cases, there may be an interval of waiting for a mail delivery in the case of certain products purchase "online." But in other cases, "cyberspace" provides a new dynamic for what Heidegger call "deseverance," that is, bringing what is remote close, that is, "within-reach." When the barriers of "lived-space" as we ordinarily understand it are "traversed," then the "vector" of closeness extends no farther than the few inches that separate one's eyes from his/her computer screen.[14] This is especially evident in the genre of Internet pornography, in which erotic images downloaded on servers across the globe are "ready to appear" on one's computer screen with the instantaneous click of a mouse.

By the same token, we cannot underestimate how advanced capitalism widens the channels for distributing the means of addiction. If in the movie, *The Graduate* (1968), the one word for the age of new prosperity is "plastics," this is doubly true, though in a different context, in the era of conspicuous consumption. Indeed, where would today's consumer be, and, indeed, the capitalistic economy as a whole, without "plastics," that is, "credit card," as "ready-to-hand," digitized extension to access

[14] For a discussion of denigration of lived-spaced into cyberspace, see Armstrong 1994, pp. 33–47.

borrowed money, i.e., debt? Where technology gives full reign to capitalism, the flip side is that the more "technical" it becomes, the more likely it is to extend the "availability" of the means of addiction. The case in point is the intersection between the forces of technology and capitalism, and the rise of a certain kinds of addiction that epitomizes the exhilaration of acquiring, possessing, and spending money, that is, addictive practices from shopping to gambling (a pursuit which crosses all economic, gender, and age divides). Loose credit, as it were, credit cards and other forms of unsecured debt (we in the United States cannot ignore the 2008 "mortgage crisis" of zero-down payment loans), creates an illusory "liquidity," which also increases the opportunity for people to gamble (not to mention shop), indeed, to "excess." A person can then gamble away money he/she does not have through proverbial "plastic" and even receive "credit" extended by such institutions as casinos. Among these new, computerized forms of "gambling" we cannot exclude "day-trading," which captivated the average investor in the late 1990s, often with dire consequences (i.e., emotionally as well financially). These examples illustrate what we earlier described as the "*commodification* of the fetish," which technology unleashes within the arena of capitalism.

While gambling was obviously prevalent in Bill Wilson's day, the "economic stimulus"—in the form of "plastic," credit cards, and lines of unsecured debt—that would elevate this activity to widespread addiction, was still decades away. Capitalism, however, has given accessibility to compulsive activities through the outlet of casinos along with the Internet. Through deregulation, capitalism not only extends the avenues of "legal" distribution for various means of addiction, but, in the process, creates a climate of tacit approval toward those various activities. We cannot overlook the role that this climate of "permissiveness" plays in licensing the activities and substances that may give rise to addiction. Perhaps cigarette smoking is one of the best examples of an addictive pursuit which society at large not only condones, but even glamorizes through the movies and celebrities that "indulge" both on and off the screen. The point of the observation, however, is not a "moralistic" one, which advocates tougher laws and prohibitions—nor even a socialistic critique of the free-market system. On the contrary, such examples point instead to the emergence of a new cultural atmosphere, which suggests the kind of paradigm shift in "life-styles" and attitudes whose roots extend back into the transformation that Heidegger envisioned with the rise of modern technology. Put simply, Heidegger argued that technology wields a double-edged sword, namely, that the greater opportunities afforded to human beings, including leisure-time, simultaneously brings its specific drawbacks and even risks. In his words, for every mode of "unconcealing" what is, i.e., the opportunities created by new innovations, there are also equally ominous modes of "*concealing*," i.e., unanticipated and destructive consequences. Indeed, much of the *danger* of technology lies in masking the latter in the former, promoting the illusion that each individual is endowed both with the critical vision and fortitude to circumvent the newest "seductions" of our age.

5.1.2 The Beginnings of a Virtual World

The 1930s ushered in the beginnings of a new technological age, as the dark cloud of a collapsing world-economy overshadowed both sides of the Atlantic (e.g., including those plunged into despair by the stock-market crash of October, 1929). As important intellectual figure in the Anglo-world during the 1930s as Heidegger was in Europe, Aldous Huxley offered a prophetic view of the direction where civilization was heading—simultaneously enhanced and ensnared by its own technological advances. In the process, Huxley addresses the problem of technology with an irony that encourages us to turn a critical eye on our culture today, as well as offers a glimpse into what dangers tomorrow may bring. When he published his epic novel, *Brave New World* in 1932, many of predictions appeared bizarre and more in the genre of what came to be called "science fiction," rather than having even the slightest ring of plausibility. Perhaps as far back as the late 1970s, before the first successful in vitro fertilization in 1978 (England), Huxley's prophecies still seemed largely farfetched. And as the prior allusion suggests, "test-tube babies" have today become commonplace (and "designer-babies" may be the next step).[15] Similar is the case with the "cloning" of animals; and the prospect of doing the same for their human counterparts, which Huxley also predicted, may be just around the corner. And, as far as issues relevant to this present study, he also foretold the rise of sexual addiction and the development of pornography on a mass scale–in the form of "interactive" participation between spectators and the display of erotic sex acts in a theater, e.g., the "feelies"–before there was *Playboy* and *Penthouse*.[16] Indeed, today's "web cams," in which the desires of voyeurism and exhibitionism converge, have made Huxley's hypothetical "feelies" a reality. Most recently, technology's outreach extends even in to one's bedroom, allowing for a web cam based, virtual sexual encounter with a pornographic actress (or actor), the development of so-called VR (virtual reality) pornography.

In the title of his book, Huxley appealed to a line in Shakespeare's play (*The Tempest*, V i), to distinguish the emergence of the future organization of a global society driven by technological advancement. The referencing to a world becomes, in Heidegger's sense, an "unworlding," in which an increasingly depersonalized and dehumanized society emerges on a global scale.[17] Of course, Heidegger did not explicitly foresee how the advance of cybernetics would lead to the creation of the Internet, in which the production of images across various multimedia platforms today mediates all social relationships. While consistent with Heidegger's account of enframing, Debord adds an important linchpin to the former's attempt to distinguish the how today's culture has been assimilated into a technological framework.

[15] The first successful in vitro fertilization procedure was performed in Oldham, England, resulting in the birth of Louise Brown, July 25, 1978.

[16] Huxley 1968, p. 152.

[17] Heidegger 1977a ("The Age of the World-Picture"), pp. 116–121, 153–154.

For Debord, the "spectacle" defines the steady stream of the arbitrary production of images, which "mediates" all social relationships (e.g., through the iconography of "texting" and "sexting"). A "pseudo-world" (in Heidegger's sense, the "unworlding") is thereby created, in which society re-emerges as the "locus of illusion and false consciousness" such that in this reign of untruth "deceit deceives itself."[18] In Heideggerian terms, the "society of the spectacle" becomes the technological persona of the "they-self," a further step in the depersonalization of everyday life. In Debord's vision of society mediated by false images, aspects of Huxley's satirical prophecy of a utopian "brave new world" have now taken now shape. In conjunction with Heidegger's account of enframing, Debord's depiction of society as refracted through the optics of technologically-produced images *indicates a deeper root from which the compulsiveness of human desires (of their fetish-like attachment) reconstitutes itself and hence from which the problem of addiction arises as well.*

The era of the Great Depression, during which Bill Wilson lived in Akron, Ohio in the 1930s, spawned great social and cultural anxiety, both in the United States and abroad. Yet, as is the case during many such times of immense turmoil, various intellectuals rose to meet the challenges of the day by wrestling with problems that would plague humankind for generations to come. Just as Bill Wilson struggled with his own addiction to alcohol, Huxley satirized the possibility of how frequently human beings would use "mood-altering" substances if all their debilitating psychological and physical properties could be removed;[19] he forecasted an impending social epidemic of addiction that would become a byproduct of the rise of modern technology. Though Bill Wilson was more or less a solitary voice in the 1930s in developing his program to treat addiction, other intellectual figures, including Aldous Huxley, would help to create a sounding board to hear his message generations later. By the twenty-first century, addiction would cease to be a problem troubling a handful of alcoholics, and instead would become so pervasive as to cut across every social class and age group. In the futuristic society of *Brave New World*, technological advancements have led to the perfection of a "drug," called Soma, which creates an equal measure of euphoria and tranquility without any negative side effects. Whenever an individual in that society finds life to be unpleasant, Soma saves the day, at least until the source of annoyance has vanished. Yet Soma's convenience and readiness for use provides a caricature to illustrate how technology begins to change the face of addiction: that is, by spurring the demand for immediate gratification, on the one hand, and, on the other, extending the availability or "within-reach" of the drugs. The pill-popping utopia of the "brave new world" is what American society *has already become* with the proliferation and availability of prescription drugs, let alone illegal narcotics.

However advanced it becomes, technology can never simply be reduced to the newest and most intricate machines. As *Brave New World* illustrates, technology involves a dramatic cultural change in the way human beings view themselves and their surrounding world, as much as anything else. As technology becomes increas-

[18] Debord 1995, pp. 12–13.

[19] Huxley 1968, pp. 36, 103.

ingly global, we view our fellow human beings more and more like the resources we use, that is, in terms of their instrumental value. For example, universities lose their ivy charm, and function more and more like corporations that process degrees for as many students as possible. But as the world becomes increasingly depersonalized, we should not be surprised that the numbers of people who become addicted, not to mention the different ways in which they do, also increases. We need only look to the rise of cases involving addiction to internet gambling or pornography, a depersonalized medium that only increases the changes of individual's looking for a "quick fix" and getting "hooked" in the process.

These examples notwithstanding, technology cannot simply be blamed as the "cause" of addiction, either. Rather, technology exaggerates the potential that human beings have to become addicted, giving fuller reign to the changes in those human capacities by which each of us may succumb to the impulsiveness of his/her cravings. Insofar as addiction involves a twofold dynamic, of change in self-perception (time) and change in the perception of the environment (space), technology accelerates the rate of these dual changes. In terms of one's self-perception, technology creates a climate of immediate gratification, modeled according to the time-frame of faster and faster internet downloads. In terms of one's perception of the environment, technology provides wider and wider access to whatever "source," substance or otherwise, which provides the opportunity to satisfy one's cravings. We cannot underestimate the importance of the change in our perception of the environment, or how we allocate space. As technological innovations make spatial distances recede, the "within-reach" of substances extends almost anywhere and everywhere. Such is the case of internet gambling, from the convenience of the laptop computer with wireless internet access. The availability of the drug is important in any scenario of addiction, and technology extends us access to ways we could hardly have imagined even two decades ago. In the case of alcohol, the convenience of access to it depends on the readily available channels for its distribution, including twenty-four hour stores and pay without cash credit cards. The use of alcohol, like cigarettes, has become so ubiquitous as to eliminate most barriers of access to it. And, as we mentioned at the outset, the development of twenty-four hour shopping, the "QVC" phenomenon, has been a dubious catalyst in spurring shopping addiction, just as casinos and the Internet has increased the opportunity to "hook" a new generation of gamblers.

While there is nothing novel about the older generation referring to the impetuousness of youth, nothing could be truer today. As in the case of the "brave new world," technology continues to produce new forms of video-auditory stimulation and arousal, including a plethora of "video-games." While during the work-day, many of us used to have very little contact with friends or family, the mobility of cell phones connects everyone almost instantaneously. As M. Scott Peck suggests, the technological climate of today greatly diminishes the importance of "delayed gratification."[20] On the contrary, immediate gratification becomes so much the norm that there is an entirely new aberration created for those unable to concentrate on

[20] Peck 1979, pp. 18–20.

lengthy tasks: "adult attention deficit disorder" (ADHD). And then, almost paradoxically, the medical profession treats the symptoms by diagnosing further "drugs," including stimulants (such as Ritalin and Adderall).[21] And so technology creates another cycle in which stimulants are required to cope with an environment of overstimulation. But few ask if we have become prisoners of technology itself, like those on a permanent "soma-holiday" in *Brave New World*.[22]

The ability to hop from one form of stimulation to another, e.g., clicking the remote-control to change from one channel to the next, plays into the hands of addiction. As such, technology does not create addiction, but perpetuates the cycle thereof by continuing to fuel the impulsiveness of the individual's cravings. The cultural climate of today creates a certain mind-set or mentality that gives greater and greater license to one's desires, without addressing the corresponding need to assume responsibility for them. The checking of one's desires, however, requires a *larger temporal horizon* within which to set priorities and seek the unfolding of a path to one's life. That temporal horizon continues to shrink within a cultural climate of "quick fixes" that views suffering as merely an obstacle to gratification. Pleasure and pain are simply servants of the one master in the name of one's obsessive-compulsive whims. Perhaps one of Huxley's characters in *Brave New World* best summarizes our contemporary mind-set of immediate gratification: "'Never put off till tomorrow the fun you can have today.'"[23] The expression of this statement in the form of a pseudo-imperative proves to be especially interesting. Indeed, the statement qualifies as an inverted version of Kant's famous "categorical imperative," which mandates putting the fulfillment of one's obligations ahead of the pursuit of pleasure. In the amoral realm of the "brave new world," hedonistic pursuits are elevated to the status of social norms. "'Take a holiday from reality whenever you like, and come back without so much as a headache or a mythology.'"[24]

When Huxley wrote his book, there was no such organization as AA, though its seeds were soon to be planted. While offering a prophetic vision of the future, *he also turned a critical spotlight on his (and Bill Wilson's time).* In one parody, a character in the novel observes: "people believed in Heaven, but all the same they used to drink enormous quantities of alcohol."[25] That is, the individuals who exist is the less than perfect society–from Bill Wilson's time until our own–have found the burden of living life to become increasingly greater as to seek alcohol as a form of escape. The desire for such escape may not be any more endemic to the "brave new world," than to Bill Wilson's or our own today. But technology has made the means much easier, as well as made the impulse toward immediate gratification much stronger. As Twerski states, "...our culture thrives on technology that eliminates waiting. We are consumers of microwaves, fax machines, cellular phones, and

---

[21] Erickson 2007. p. 101.

[22] Huxley 1968, p. 103.

[23] Huxley 1968, p. 62.

[24] Huxley 1968, p. 36.

[25] Huxley 1968, p. 35.

instant foods. Even if one can envision 'happiness' coming later in life, the prevailing ethos of instant gratification makes a long wait intolerable."[26]

Yet the constant irony in all of this, which the inhabitants of our global society today share as a possible precursor to a "brave new world," is that people overindulge in alcohol both in the exceptionally happy and the unusually sad times.

5.2 The Onslaught of Boredom and the Addictive Search for Euphoria

Ironically, the more people are bombarded by stimuli, the more easily they can succumb to periods of idleness and indifference, in short, to boredom. But boredom is not just a state of profound indifference. It is always a state in which an individual finds his/her time to be meaningless, in such a way that no activity or pursuit can emerge as vital. On the surface, boredom appears as the opposite mode of awareness to the hustle and bustle of technology. Yet understood more fundamentally, as what Heidegger calls a prevailing "attunement" (*Stimmung*), rather than just a subjective state or attitude, boredom exhibits a different character. For boredom outlines the encompassing backdrop out of which we experience beings *in* a whole (*Seiende im Ganzen*), insofar as everything recedes into a veil of indifference. As a result, boredom points to a preliminary disclosure of the mundaneness of our everyday situation which impels us, as if seeking an "escape," to shift our attention to the "excitement" provoked by individual things. In the case of technology, where the "instant" offers the greatest fascination, the excitement of such mundane activities, e.g., "computerized war-games" (e.g., as a special genre of video-games), becomes especially pronounced. Although experienced as a possible escape from boredom, the excitement and allure of technology still confirms the power of indifference as casting its spell over everything. As Heidegger emphasizes, the danger does not lie with boredom as such, but, rather the reverse, in never pausing to question what this pervasive indifference signifies: namely, the ease by which humanity can slip into the pursuit of "meaninglessness" or the vacuum created with a loss of any sense of "meaning" and purpose to life:

> This *profound boredom* is the fundamental attunement. We pass the *time*, in order to master it, because time becomes long in boredom. Times becomes long for us. It is supposed to be short, then? Does not each of us wish for a truly long time for ourselves? And whenever it does becomes long for us, we pass the time and ward off its becoming long! Boredom, long time: especially in Alemannic usage, it is no accident that 'to have long time' means the same as 'to be homesick.'[27]

If a sense of "loss of home" distinguishes the awareness of boredom, then the behavior by which an individual seeks "escape" in the newest diversion provides an

[26] Twerski 1997, p. 14.

[27] Heidegger 1983/1995, p. 120; tr. 80. For further discussion see Emad 1985, pp. 63–78.

important signpost. Specifically, this tendency toward escape, as it were, escalates to the level of an "addiction," then the addictive behavior suggests a kind of "uprootedness." The uprootedness describes a common way we come to experience our isolation and alienation, a loss of a sense of our "place" in life, a lack of connectedness to others and community. As such, boredom and a sense of homelessness occur together to indicate the malaise and sickness of our age. The more one is aware of what one is doing, rather than just doing it, the more one notices the length of the span of that activity. That is, time, turns in the direction of a future that does not appear to offer anything exciting, thereby leaving the present empty of any specific focus of interest. The addict, however, finds this emptiness of inactivity to be especially disconcerting, because it stands in stark contrast to his/her desire for immediate gratification. The threat that boredom can become overwhelming and stifling poses one of the greatest threats to the addict's being able to relinquish the lure of his/her cravings.

While boredom appears to be negative to most of us, it also serves as a reminder of something very important to the potential well-being of the addict: a sense of calm and repose. The ability to reside in this calmness has an important therapeutic value. The problem with addiction is that it plunges the addict into a life of turmoil and confusion. Equally as troubling, the addict's lack of concern for everything else but satisfying his/her cravings tears at the fabric of others' lives, replacing order with chaos. The more chaotic the addict's environment becomes, the more volatile his/her emotions become as well. Strangely enough, boredom becomes like a buffer to this emotional volatility. When the buffer breaks, a flood of emotions arises which, in the worse case scenarios, can translate into destructive actions: from fits of rage to domestic violence. In a positive manner, boredom can provide a temporary haven from the turmoil of the emotions. But the emotions are themselves like pockets of energy, and must find an outlet of expression in one guise or another. When measured against a spectrum of calm and repose, the addict faces the specific challenge of managing his/her emotions.

While boredom implies monotony, the addict relishes "change," at least as far he/she experiences the pleasantness or unpleasantness of any given situation. Today we use the term "mood-altering" to describe the desired state of *change* that the "quick-fix" of the substance provides. But here's the rub. All human beings are prone to moods of one kind or another. We may wake up in a good or bad mood, or be elated or downcast given the turn of certain events. What is unusual is relying upon something external, specifically a substance to interact with our bodies in order to induce drastic "mood-changes." A natural, and perhaps healthier way of altering one's mood, would be exercise and other forms of physical recreation. But an artificially induced way of altering moods, corresponding to the advancement of technology, proves to be more problematic. For in the former case, the individual accepts the challenge of embracing the complexity of one's emotion; in the latter case, one seeks an avenue of escape. Better to change the emotions than to face them! Or, as a contemporary aphorism states: "Medicate rather than meditate."

Thus, the question arises as to development of different instruments to induce this "mood-altering" effect, thereby the manufacturing of synthetic instruments to achieve the desired end. Even if only in so-called primitive form, these synthetic instruments still fit within the overall blueprint of production and consumption, developed through the medium of the artifice itself as commodity to be bought and sold. Whether in a pharmaceutical form (e.g., synthetic opiates) on in the derivative guise of a street drug (e.g., crystal meth), the artificial production and supply of instruments to meet a specific demand of self-indulgence only reveals the intersection between modern technology and the commodification of diverse, potentially addictive instruments. Technology encompasses even the drug industry, which in turn takes shape and is mobilized through the *gestalt* of the artifice. As various industries, both legal and illegal become wedded to this *gestalt*, the corresponding temporal span of immediate gratification (and spatial deseverance of the "within-reach" of availability) dictate, conversely, that the degree of gratification (e.g., of "mood-alteration") is equally brief or specious.

As Heidegger realized, boredom is a multifaceted phenomenon, insofar as it indicates our ambiguous stance toward technology on the one hand, and, on the other, reveals how our immersion in it opens new avenues by which we can become addicted. Indeed, the more pervasive and deep the boredom is, the more prone human beings are to seek out, and become captivated, by the newest fads and diversions. We might pose the question this way: Is it merely an accident that, with the rise of leisure and affluence throughout the world, there has also been a multiplication of kinds of addiction? Or stated another way, in an earlier historical period when the time allotted to working and thereby surviving were paramount, and, accordingly, there were fewer hours allocated to leisure, was the opportunity or inkling to become addicted nearly as strong?

In asking this question, we bring to the forefront two key issues pertaining to the changing face of addiction within the technological age. First, technology makes our experience of time appear especially fluid. That is, the management of time through the calculation of smaller units makes us more readily unsettled by the preponderance of lengthy intervals of inactivity (e.g., the prevalence of boredom), on the one hand, and, on the other, more easily tempted by the expectation of various forms of immediate gratification. Second, technology alters the ebb and flow of everyday time, including the division between night and day, disrupting if not modifying natural rhythms. For example, the more we seek to "get done" by compressing time into various activities, the more we seem to "leave undone." Indeed, despite all the convenience and "time that is saved" by the latest gadgets, we seem to have less and less time to unwind and actually sleep. Within the modification of everyday time, the excitement of the instant is overshadowed only by the tedium of routine. In this way, the allure of instant pleasure and the monotony of boredom become two sides of the same coin; they are both motivations toward addiction.

Technological progress certainly leads to greater conveniences, like ATM machines, if not greater material comfort. No doubt this progress cannot be separated from the affluence which it helps to create, the "wealth-effect" in which more households in America qualify as "millionaires" than even before. But with increasing affluence and leisure also come opportunities for alcohol and drug abuse, not to mention a fascination with "artificial" stimuli like online pornography and gambling.

Drug use is as much a "game" for Beverly Hills celebrities, as it once was for those trapped in the ghetto. Perhaps lost on many of us is that just as technology makes the "world flat," it also grants "equal opportunity" as far as the chance of becoming addicted goes.

5.3 Cultural Excess and the Illusion of a "Cure"

With the rise in affluence and its numerous comforts, there is also a decrease in the interest in spirituality. The Twelve Step Program that Bill Wilson pioneered is first and foremost an exercise in spirituality, and perhaps even an "experiment," insofar as it is not predicated on any set of orthodox beliefs. But as the spiritual vacuum of our age increases, and addictions to various substance and activities become more common, how effective can the Twelve Step Program continue to be? Indeed, those within the medical community have come to advocate physiologically based treatments, new forms of medication and "prescribed" drugs to combat the neurological reactions of those with various addictions and obsessive-compulsive disorders.[28] In effect, these are more sophisticated versions of what the medical community first developed to treat alcoholism, specifically, "Antabuse."[29] When combined with the consumption of alcohol, Antabuse creates such horrific physiological reactions as to serve as a "deterrent" to drinking altogether. Yet even with the latest insight in the work of our brain chemistry, will the invention of new drugs finally solve the problem of addiction, or, on the contrary, reinforce *the illusion of technological supremacy in the "brave new world,"* namely, that a mixture of prescription drugs can solve problems that in truth speak from the depths of the human predicament?

Are we then on the threshold where the drug companies replace churches as the new havens of salvation, and the physicians become the new prophets? So long as it does not disappoint, technology courts its own congregation of believers. Occasionally, technology does fall short, for example, when medicine cannot help the terminally ill, or even in cases of terrorism where commercial airliners serve as unlikely missiles. More than a half century before Freud, Karl Marx hypothesized that the development of material security on a global basis would make religion obsolete. Freud also believed that psychological maturation would turn spirituality into a useless crutch. But neither anticipated that a society predicated on affluence and happiness creates its own culture of obsessive-compulsiveness. Thus, if religion ceases to be the "opium of (escape for) the people," it is not simply because spirituality has become obsolete. Rather, technology has given the individual a license to

---

[28] Erickson 2007. pp. xi–xiii. From a parallel view of neuroscience, Nicholas Karadas points to how the brains scans of children who are "hooked" on electronics mirror the brains scans of those people who are addicted to opiates. This provides an interesting medical "fact" as to (1) the overlaps between "behavorial" and "substance" addictions and (2) how the technology of "enframing" provides a new undergird for the problem of addiction. See Kardaras 2016, pp. 1–13.

[29] Erickson 2007, pp. 166–167. Also see Steiner 1971, pp. 114–115.

pursue his/her cravings, in such a way as to make synthetic opiates, e.g., prescription painkillers, literally the "new opium of the people."

The fact that so many individuals within society, least of all addict, admit to being "stressed out" only confirms why addictive disorders have become so pervasive. In the case of alcoholism in general, and other addictive disorders in particular, we cannot help but be struck by the proximity and breadth of the problem. As in the case of cancer, most of us know someone who has been afflicted by the problem of alcoholism. Addictive tendencies may begin innocently enough as a social and recreational activity, and then, under so-called conditions of "stress," escalate into a dangerous form of escape from the vicissitudes of life. Such popular expressions as being "stressed out," more than just emotional venting, may also serve as "indications" for interpreting the ineffectiveness of many historical attempts to counteract the increase of addictions. Indeed, these well-intentioned "panaceas" speak from the misconceptions of our culture that actually empower the addiction, rather than the individual who seeks to break its grip. If not strictly forms of "institutionalized" co-dependency, they arise within the same orbit of misunderstanding that pervades the addictive culture of "excess" itself. For example, the just say "no" campaign of the 1980s, with Nancy Regan as its primary spokesperson, supposedly took the moral highroad but with little or no positive effects: "Just say 'no' to smoking, drinking, sex, drugs, and rock and roll," summarizes the message of that campaign. But, as these pundits fail to see, the culture of "excess" cannot be reduced simply to various negative behaviors, for the climate of conspicuous consumption encompasses both the moral zealots and the addicts. We need also recall that the acceptance of cocaine as the preferred drug of recreational use among the most affluent classes in the United States during the 1980s directly parallels the surge of the great "bull market." Likewise, the "temperance" movement of Prohibition in the 1920s and early1930s also floundered on the mistaken presumption that moral values can be "taught" in order to check the individual's tendency to over consume alcohol and give into his/her cravings to drink.[30]

The pursuit of temperance, however, is not necessarily a bad idea, provided we appreciate its deeper implications. But temperance is not simply about managing the amount of one's intake of alcohol or any given substance. Rather, it is about the management and regulation of oneself as an emotional being. It was over two thousand years ago when Aristotle proposed the idea of the "mean," as the middle ground between "excess and deficiency." From him, and his teacher, Plato, arises the moral ideal of *moderation*. First and foremost moderation involves avoiding the emotional extremes that makes the individual prone to self-destructive actions towards him/herself, as well as harming others. But the "balance" that one seeks does not begin as an "external" prohibition, but instead originates as a form of discipline that can only be administered "internally." If Aristotle were alive today, however, his message would not simply be to live a life of temperance and "moderate" one's use of alcohol. All that is well and good, and perhaps even self-evident; and yet, as in the case of Nancy Reagan's plea, may fall on deaf ears. The question of temperance,

[30] Kurtz 1991, p. 229.

then, begins with the handling of one's emotions, without giving in to their chaotic impulses or their tendency to vacillate to extremes. The case in point is the individual, on the spur of an emotion, seeks immediately to annul it by reaching for a drink or popping a prescription tranquilizer. The tendency toward over indulgence has its corollary, but not necessarily its cause, in this aspect of the personality. And here our assessment of this aspect of addiction becomes murky. For personality tendencies and traits do not always translate into a specific behaviors like addiction. By the same token, there may be addicts exhibiting a distinctive behavior who may not fit into the mold of an addictive personality.

The fact that addicts seek the euphoria of "highs," which may in turn be followed by "lows," already provides a sign, perhaps a symptom, of an emotional tendency. But is a vacillation between emotional extremes already inherent in that search? Is addiction, emotionally speaking, a case of "nature," "nurture," or both? Perhaps there is no simple answer to this question. Whatever its origin, the addict's personal experience of being addicted can best be summarized by Bill Wilson's observation that alcoholics are "'all or nothing people.'"[31] This means that alcoholics not only ignore limits, but that they view life self-destructively as offering only extremes between perfection and resignation. That is, only the realization of the ideal, rather than its partial fulfillment or approximation, will suffice.

Let us take an example outside the framework of the addict. I have a student who wants to apply to graduate school, among a pool of applicants who are equally competitive. The reality of the situation would, in terms of the student's approach, suggest a "prudent" strategy: apply to various schools, of greater and lesser prestige and an inversely lower or higher chance of admittance. But more frequently than one might expect prudence does not win out. On the contrary, students often reject this strategy in the name of an "all or nothing" approach: "If I don't get accepted for graduate studies at Yale or Princeton, then I am not going anywhere!" An appreciation of *prudence*, which Aristotle understood as "balanced judgment" and from which our judicial system derives its concept of "jurisprudence," goes completely out the window.[32].

This facet of "addictive thinking" has been characterized in terms of an "either-or" perspective.[33] The hallmark of this perspective, however, lies instead in the weight of the negative, that is, the grammatical form of "neither-nor": "If I *don't* get into Yale....then I am *not* going anywhere." The difficulty lies not only in the individual's (in this case, the student's) unrealistically "high" expectations, but rather in his/her summarily dismissing an intermediary level of fulfillment. And it is the negative spirit of this disavowal, which becomes the origin of a problem. Not only does the individual's fallacious thinking lead him/her down a path of inevitable frustration. But the individual holds on to the frustration, which in turn can translate into a negativity of low esteem, a sense of failure, and even self-loathing. While the

[31] Kurtz 1991, p. 229.

[32] For a discussion of the "mean," see Aristotle 1998, pp. 43–47. For a discussion of Aristotle's impact on a Heideggerian ethic, see Hatab 2000, pp. 103–112.

[33] See Twerski 1997, pp. 54–55.

"perfectionist" attitude poses a difficulty, the larger problem springs from the "negative" way of coping with imperfection, that is, by exaggerating it through the positing of false disjunctions: "neither-nor," if *not* all, then *nothing*.

Unfortunately, "nothing" is an unfavorable alternative, and the life of the addict is evidence of this. To a certain degree, the "nothing alternative" places the individual in an impossible situation, putting him/her at odds with even his/her own best interests. As the old adage goes, "we have met the enemy, and it is us." This is particularly true in the case of the addict, whose worst enemy is him/herself. Not only does the "all or nothing" thinking spark the emotional turmoil that makes the consumption of a substance appear as an attractive substitute for happiness. But that thinking also hinders any chance of recovery. For these negative thought-patterns make it appear "illogical" that recovery can begin with a single step, followed by equally modest ones. Indeed, modesty implies "moderation," which is precisely the mentality, both in thought and action, which the addict rejects from the outset! Moreover, our culture of excess encourages individuals to seek "extremes," e.g., of wealth, passion, and pleasure over and against any measure of moderation.

The "all or nothing" mentality does not grow up in a cultural vacuum,[34] even if it distinguishes those members of society who are most prone toward addiction. Perhaps of all the addictions, gambling epitomizes this mentality, while providing a clue as to corresponding tendency within our culture: its preoccupation with what prosperity and all the spoils that come from succeeding in a capitalist society. Capitalism not only emphasizes prosperity, but predicates success on achieving a competitive edge at the expense of others, whether individually or in the market place. To be sure, the tendency for the individual to become addicted does not hinge on any economic or political system. After all, the lure of vodka, even in Dostoevsky's time, was enormous within the pre-industrialized society of Russia and the subsequent communist regime of the Soviet Union. In the case of capitalism, however, the cultural icon of being "number one" closely parallels the "all or nothing" mentality of the addict.

There is an old adage in sports that no one remembers the person who "finishes second." Famous slogans such as "winning is not everything, it is the only thing," illustrate how strongly the "number one" mind-set has become entrenched in the everyday American's vision of success. Indeed, the *logos* of such a statement reflects the *mythos* of the "American dream." A peculiar nexus exists between society and sports, whereby the capitalistic desire to accumulate great wealth outstrips the "love of the game," as in the spirit of the etymology of the word "amateur." But if professional athletics is a microcosm of American society, then betting on sporting events becomes a metaphor for how *various cultural customs and practices converge to engender the subliminal patterns of addictive thinking*. We speak of "subliminal" to underscore that these thought patterns arise unconsciously. But precisely because of this, the "all or nothing" mentality becomes especially intense. What else can be more electrifying than betting all your money on a singular sporting event like the

[34] See Romanyshyn and Whalen 1987, pp. 203–207 ("Franklin and Gatsby: Archetypal Self-Made Men"). I am grateful to Michael E. Zimmerman for providing me with this reference.

Super Bowl in which all the competitors are committed to the "winning is the only thing" philosophy?

Particularly with the convenience that the Internet provides, it is not surprising that gambling has become one of the faster growing forms of addiction. Moreover, how can a society, which preaches being "number one" at all costs, also foster as its aberrant byproduct the "all or nothing" mentality of the addict? We cannot ignore the obvious parallels that exist between many noteworthy cultural practices and the psychological dynamics of addiction. As inherently self-destructive as this all or nothing may be, its tendency toward "perfectionism" harbors a curious paradox. The addict, as we have seen, may very well be the individual who aspires to the most exalted heights, a harbinger of the "spirit." As Jung emphasizes, the flip side of indulging in spirits is being a "spirited" individual, that is, one capable of the greatest foresight and vision. A contemporary Jungian, Billye B. Currie, has claimed further, that in the specific case of gambling, there is a "primitive" desire, e.g., in the name of "lady luck," to court the favor of the "gods" and thereby ascend to a higher spiritual plane.[35] Bruce Wilshire develops a parallel argument in suggesting that that "ecstasy deprivation" is a key trigger in the penchant for the individual's becoming addicted within contemporary society.[36]

When developed in a realistic and practical manner, the drive toward perfectionism and nascent spirituality of the addict may also be a gift. But, when perverted into the negativity of the "all or nothing mentality," that perfectionist attitude may easily translate into a curse. In its worst form, the curse may become a festering self-loathing, which overtakes all optimism and joy in life. Perhaps the fear of this self-loathing can explain why isolation can be extremely destructive for the addict: because left to its own resources, an isolated mind can be a very threatening place in which to spend the majority of one's time.

In a book published decades after *Brave New World*, Aldous Huxley examines the link between spirituality and drugs, seeking a common denominator:

> The urge to transcend the self-conscious self is, as I have said, a principal appetite of the soul. When, for whatever reason, men and women fail to transcend themselves by means of worship, good works and spiritual exercises, they are apt to resort to religion's chemical surrogates–alcohol and "goof pills" in the modern West, alcohol and opium in the East, hashish in the Mohammedan world, alcohol and marijuana in Central America, alcohol and cocoa in the Andes, alcohol and the barbiturates in the more up-to-date regions of South America.[37]

Huxley's remark closely resembles Jung's suggestion in his letter to Bill Wilson, namely, that an unfulfilled, "higher" aspiration often harbors a hidden motivation in an individual's turn to drugs and alcohol. With the marketing of the "illusion" of fulfillment and happiness on global scale, on the one hand, and the

[35] Currie 2001, pp. 17–22. For one of the first literary presentations of the addictive character of gambling, see Dostoevsky 1968, pp. 448–451. Dostoevsky provides a poignant, yet comical portrayal of a Granny transfixed at the roulette table.

[36] Wilshire 2003, pp. 3–7.

[37] Huxley 1954, p. 67.

commercialization of products holding such allure, attraction, and promise, on the other, perhaps at no time in human history have vast proportions of the population been put at risk to becoming addicted to various substances and/or activities.[38] But whatever personal allure various commercialized products may pose, their "artificial" origin reveals a deeper motivation behind an individual's becoming addicted than might otherwise be attributed to an individual's hedonistic impulses becomes more pervasive, so does the imposition of technological forces extend into every corner of life. But the ubiquity and impersonalizing of such control only occurs at the expense of depriving the individual of any claim of self-mastery. The relegation of the individual to a mere cog in the corporate machinery maximizes the controlling mechanisms of enframing, precisely by rendering the individual powerless and incapable of exercising control over his/her own existence. The more acutely the individual experiences this sense of fragmentation and powerlessness, the wider is the vacuum (including of loss, despair, and meaninglessness) which is created for the self to seek substitutes, surrogates, and "artificial" means for regaining control (no matter how false and illusory).

On the cusp of modernity, as Heidegger suggests, man becomes the "laboring animal" (*arbeiterden Tier*).[39] Humanity's worth is defined by its capacity to produce, particularly on a mass scale.[40] Under the sway of technology, man re-emerges as the "control-obsessed" animal, whose compulsive cravings point back to a deeper root from which the problem of addiction emerges *as* a problem.[41] In this way, technology reveals the two-sided (Janus-character) of its ambiguity by exacting complicity from the individual. In occupying dual roles as producer and consumer, the individual succumbs to the *allure of the artifice, which as a technical mediator of human desire provides a vehicle to regain control otherwise lost in the end game of enframing.*[42] In the specific instance of *an* addiction, the allure of "chasing the high"[43] is equally regaining the euphoria of that experience, by inauthentically seeking to recapture the past so to bring into the (inauthentic) future the *capacity to control* the present. Kemp is thereby accurate in suggesting that the self's fixation on the immediacy of the 'now' "evolves out of prototypical experience of the 'high'".[44] In this inauthentic, unowned temporalizing, regaining control is still name of the game, as the precondition for sustaining the feeling of euphoria. In intensified falling, the untruth or illusion of achieving control in this way, is then carried out on

---

[38] For an important discussion of the development of different kinds of addiction cross- culturally, see Alexander 2010, pp. 1–6.

[39] Heidegger, "Überwindung der Metaphysik," 2000b/1973, p. 71; tr. 86. Also see Heidegger 1998, "On the Question of Being," pp. 291–301. Heidegger addresses Ernst Jünger's vision of the mobilization of work, and its connection to a "planetary" or "global" outlook of modern society.

[40] One of Heidegger's students, Herbert Marcuse, makes this point in emphasizing the difference between "work" and "play". Marcuse 1955, pp. 190–195. For further discussion, see Kovel 1987, pp. 330–348.

[41] Zimmerman 1995, p. 501. Also see Schalow 2006, pp. 161–167.

[42] Schalow 2003, p. 27.

[43] Pinsky 2003, p. 143.

[44] Kemp 2009, pp. 1–18 (esp. p. 15).

a second level as being "out of control," or losing control over one's ability to choose and offset the overwhelming power of the addiction. In this addictive phase, the self's vacillation between the extremes of seeking and losing control mirrors the illusion, the *hybris*, that technology will allow us to master our destiny despite abandoning us to the totalizing drive of enframing.

In a similar way to Nietzsche's eternal recurrence,[45] technology draws humanity into a cycle of *depending upon artificial means to supply a false sense of self-control, in order to reap immediate benefits only through the complicity of ignoring the detrimental side effects.* Technology contributes to this cycle of dependency. In this cycle, the individual's addictive *reliance* upon the commodification of specific products and their fetish-like promise of immediate gratification become a mirror image of each other (e.g., the earlier marketing of cigarettes to women as "glamorous" in the 1960s). We thereby come to the important realization that technology creates a cycle of dependency through which addiction can flourish as a global epidemic, and, conversely, that the phenomenon of addiction serves a microcosm to reflect the larger, global issues at work in today's technological world. Thus, the story of addiction is more than only the compulsion experienced by the individual who is addicted, but instead includes a deeper subtext of our encounter with the Janus-faced character of cyber technology: that is, our control based drive for self-mastery makes us increasingly vulnerable to the capriciousness of our desires, whims, and urges.

If this conclusion proves to be true, then any attempt to confront the problem of addiction, and seek avenues for its treatment, must proceed by addressing the macro issues embedded in the way that cyber technology provides a ubiquitous platform to mediate human desires. By the same token, the more globalized our community becomes, the more vulnerable we become to the onslaught of its various trends, including secularization, urbanization, and depersonalization. Within this extensive climate of distress and boredom, the individual's opportunities for self-discovering becoming even more restricted. Traditional religious and educational institutions becoming increasingly co-opted by technological forces, from pressures of marketing to the media. A kind of spiritual vacuum which traditional religions and denominations have had a difficult time filling. In the twenty-first century, the Twelve Step Program offers a formula (within an otherwise secular setting) to rekindle an interest in spirituality when other more "traditional" approaches flounder. The Twelve Step Program has the distinct advantage of transmitting a message about God that can speak across various cultural, economic, and intellectual divides. And the message carries such a universal reach, precisely because Bill Wilson proved to be somewhat of a "prophet" in crystallizing Christianity into its most basic elements.

Human beings have perennially searched for some measure of serenity in their lives. Today, that search proceeds in as many "unorthodox" channels, from yoga to "New Age" mysticism, as orthodox ones. The more human beings find their lives to be fragmented within technological society, the more they seek spiritual solutions to their questions. The Twelve Step Program occupies a unique place within the

[45] See de Gennaro 2012, pp. 201–233. Within a technological framework, the eternal recurrence is understood as a vehicle for the will to power as the relentless drive toward the imposition of control and mastery over all that is.

pantheon of these religious movements, blending elements new and old. And this is as good an explanation as any as to why the Twelve Step Program continues to have the impact it does today. Yet, within the age of "excess," we must wonder whether that program has outlived much of its effectiveness.

5.4 Conclusion

In our pursuit of material comforts, are we any more superficial than those of the past? Has technology transformed us into a more efficient, more prosperous people, albeit shallower as well? Have the solid virtues of personal responsibility and hard work gone by the way side in favor of a culture of "conspicuous consumption" and the "greed is good" credo of the movie, "Wall Street"? Have we become victims of another wave of narcissism, in which the immediate availability of a "selfie" epitomizes a new technological genre of immediate gratification? It is certainly problematic for any culture to cast moral dispersions on any other. Yet our age of prosperity rewards individuals less for their acts of virtue than for their marketable skills, for example, in entertainment and athletics. Even virtue as a purely moral category going back to Aristotle must be extended to include more broadly the manner of our abiding within an *ethos* that can foster a openness beyond the rule of instrumentality of enframing.[46] But perhaps more disturbing is how much dissimulation and excusing-making—the cornerstones of co-dependency (as outlined in the previous chapter)—occurs in our everyday interactions with others. And technological conveniences like the cell phone have reinforced deceptive tendencies, e.g., saying one is calling from home to report in sick, while really out shopping. Are we more vulnerable today to lying, from our top political officials on down? (Do we need to look any further to the 2016 Presidential Election in the United States?) While some moralists might say "yes," the problem may be particularly acute in a culture that seems to condone, and thereby perhaps "enable," the self-deceptive practices that sustain addiction, e.g., lying and dissimulation.

As the enduring spiritual plan of our today, does the Twelve Step Program still speak to this inherently problematic mind-set of the addict, from deception to denial, from excuse-making to rationalization? Perhaps the greatest irony here is that the mind-set of the addict, both individually and culturally, lies in overcoming the expectation that any program of treatment can be tailored to conform to one's time-frame of immediate gratification. On the contrary, the interval of treatment is long, protracted, and ultimately for the rest of one's life, which clashes with the addict's own time-frame of always seeking immediate gratification (not to mention relishing in a new cultural mind-set of narcissism). The so-called danger of technology lies not only in making us increasingly dependent upon it, but in mediating human desires through the lens of immediate gratification. Thus, the hand-held device becomes a contemporary cultural stamp of approval upon the self's capitulating to

[46] For further discussion see, Kirkham 2016, pp. 97–119.

its impulses and indulgences. *Technology then becomes a chief way, through its manner of "enframing," of channeling our desires into the instant "time-frame" of immediate gratification.*

Through "enframing," does the rise of the artifice conceal a tendency for human beings to insulate themselves from the *pathos* of the human predicament, and thereby make them more vulnerable to their own addictive tendencies? Our effort to unravel this paradox yields a direction not only for re-examining the problem of addiction *as* a problem, but also for developing a critical perspective to assess different plans of treatment. Given this vantage point, let us undertake an archaeological explication of the religious and philosophical roots of the first plan of treatment, the Twelve Step Program.

References

Aho, Kevin. 2009. *Heidegger's neglect of the body*. Albany: SUNY Press.

Alexander, Bruce K. 2010. *Globalisation of addiction: A study in the poverty of spirit*. Oxford: Oxford University Press.

Aristotle. 1998. *Nicomachean ethics*, trans. David Ross. Oxford: Oxford University Press.

Armstrong, William. 1994. Cyberspace and the relation between being and place. *Southwest Philosophy Review* 10 (2): 33–47.

Currie, Billie B. 2001. *The gambler: Romancing lady luck*. New York: Inner City Books.

de Gennaro, Ivo. 2012. Nietzsche: Value and the economy of the will to power. In *Sources and readings on a key concept of the globalized world*, ed. Ivo de Gennaro, 201–233. Leiden & Boston: Brill.

Debord, Guy. 1995. *Society of the spectacle*. New York: Zone Books.

Dostoevsky, Fyodor. 1968. *The gambler*, trans. Constance Garnett, in *Great short works of Fyodor Dostoevsky*. New York: Harper & Row.

Emad, Parvis. 1985. Boredom as limit and disposition. *Heidegger Studies* 1: 63–78.

———. 2013. *"History" and "nothingness" in Heidegger and Nietzsche*. Budapest: Societas Philosophia Classica.

Erickson, Carlton K. 2007. *The science of addiction*. New York: W. W. Norton.

Hatab, Lawrence J. 2000. *Ethics and finitude: Heideggerian contributions to moral philosophy*. Lanham: Rowman and Littlefield.

Heidegger, Martin. 1973. Overcoming metaphysics, in *The end of philosophy*, trans. Joan Stambaugh. New York: Harper & Row.

———. 1977a. The question concerning technology and the age of the world-picture. In *The question concerning technology and other essays*, trans. William Lovitt. New York: Harper & Row.

———. 1983. *Die grundbegriffe der metaphysik*, GA 29/30. Frankfurt am Main: Vittorio Klostermann.

———. 1989. *Beitrage zur philosophie (Vom Ereignis), GA 65*. Vittorio Klostermann: Frankfurt am Main.

———. 1995. *Fundamental concepts of metaphysics*, trans. Nicholas Walker and William McNeill. Bloomington: Indiana University Press.

———. 1998. On the question of being, trans. William McNeill. In *Pathmarks*, ed. William McNeill. Cambridge: Cambridge University Press.

———. 1999. Contributions to philosophy (From Enowning), trans. Parvis Emad and Kenneth Maly. Bloomington: Indiana University Press.

———. 2000b. Die frage nach der technik and Überwindung der metaphysik. In *Vorträge und aufsätze*, GA 7. Frankfurt am Main: Vittorio Klostermann.
———. 2003. *Four seminars*, trans. A. Mitchell and R. Raffoul. Bloomington: Indiana University Press.
———. 2012. *Bremen and Freiburg lectures: Insight into that which is and Basic principles of thinking*, trans. Andrew J. Mitchell. Bloomington: Indiana University Press.
Huxley, Aldous. 1954. *The doors of perception*. New York: Harper & Row.
———. 1968. *Brave new world*. New York: Harper & Row.
Idhe, Don. 2010. *Heidegger's technologies: Postphenomenological perspectives*. Bronx: Fordham University Press.
Jung, C. G. 1965. *Memories, dreams, reflections*, trans. Richard and Clara Winston. New York: Random House, Inc.
Kardaras, Nicholas. 2016. *Glow kids: How screen addiction is hijacking our kids and how to break the trance*. New York: St. Martin's Press.
Kemp, Ryan. 2009. The temporal dimension of addiction. *Journal of Phenomenological Psychology* 40 (1): 1–18.
Kirkham, Nin. 2016. Recognizing our place in the world. *Environmental Ethics* 38 (1): 97–119.
Kovel, Joel. 1987. Schizophrenic being and technocratic society. In *Pathologies of the self*, ed. Michael David Levin, 330–348. New York: New York University Press.
Kurtz, Ernest. 1991. *Not-God: A history of alcoholics anonymous*. Center City: Hazelden.
Marcuse, Herbert. 1955. *Eros and civilization*. New York: Random House.
Peck, Scott M. 1979. *The road less traveled*. New York: Simon & Schuster.
Pinsky, Drew. 2003. *Cracked: Putting broken lives back together*. New York: Harper-Collins.
Polt, Richard. 2015. A Heideggerian critique of cyberbeing. In *Horizons of authenticity in phenomenology, existentialism, and moral psychology: Essays in honor of Charles Guignon*, ed. Pedersen Hans and Altman Megan, 179–197. Dordrecht: Springer.
Romanyshyn, Robert B., and Brian J. Whalen. 1987. Depression and the American dream: The struggle with home. In *Pathologies of the modern self*, ed. David Michael Levin. New York: New York University Press.
Schalow, Frank. 2003. Everydayness and the question of human addiction: A phenomenological approach. *Southwest Philosophy Review* 19 (2): 17–30.
———. 2006. *The incarnality of being: The earth, animals, and the body in Heidegger's thought*. Albany: SUNY Press.
Steiner, Claude. 1971. *Games alcoholics play*. New York: Grove Press, Inc.
Twerski, Abraham. 1997. *Addictive thinking*. Center City: Hazelden.
Wilshire, Bruce. 2003. *Wild hunger. The primal roots of modern addiction*. Lanham: Rowman and Littlefield.
Zimmerman, Michael E. 1990. *Heidegger's confrontation with modernity: Technology, politics, art*. Bloomington: Indiana University.
———. 1995. Ontical craving versus ontological desire. In *From phenomenonlogy, to thought, desire, and errancy: Essays in honor of William J. Richardson*, ed. Babette Babich. Dordrecht: Kluwer Academic Publishers.

Part II
From *Pathos* to Healing: Transcendence and a New Life-Trajectory

Chapter 6
From Theology to Therapy: A Genealogical Account

The phenomenon of addiction no longer distinguishes only an aberration or affliction, a malady affecting a small subset of society. On the contrary, through its diagnosis, the problem of addiction also *reveals* a larger cultural crisis: that is, the *danger* of our increasingly dependence upon technology. But what does this game-changing development mean for the ongoing search for a treatment to this vexing problem? In this chapter, we will raise this important question, and consider whether there is a further, Janus-faced or two-side character of technology that is in play, which masks the extent of our dependency on technology and hence how deep seated the fetish-like craving of our desires may be.

In retrospect, then, the rise of the culture of conspicuous consumption and the emergence (and diagnosis) of addiction *as* a problem are not coincidental. Instead, these two developments occur in tandem, insofar as the fulfillment of certain levels of material comfort also create the illusion of happiness, well-being, and even spiritual solace. Perhaps it is not surprising, then, that the first program for treating addiction would arise from a religious movement, rather than, as we might expect in the case of a medically based illness, from the adjacent field of science. Yet, religion in its various permutation speaks to the deepest recesses of emotional conflict, which those who succumb to addiction experience to an inordinate degree: *a forlornness, hopelessness, and despair that begged for a spiritual as well as a clinical and medical resolution.*

How can a religious movement, loosely connected with Christianity (broadly speaking, the "Oxford Movement"), spawn a therapeutic practice for treating addiction? In posing this question, we must recall that such a distinction was not readily available in the early 1930s when this first treatment program was developed in the United States; indeed, matters were not as clear cut as they might be today in our increasingly secularized culture. Rather, those needing some kind of relief from their malady, and seeking new direction for their lives, sought insight if not answers anywhere and everywhere. As Plato once stated in The *Republic* nearly 2500 years

F. Schalow, *Toward a Phenomenology of Addiction: Embodiment, Technology, Transcendence*, Contributions To Phenomenology 93,
DOI 10.1007/978-3-319-66942-7_6

ago, "Necessity is the mother of invention."[1] Indeed, what came to be called the "Twelve Step" Program is as much one part necessity, one part inspiration. The inspiration emerged through an appeal to various religious sources, from which the pioneers of the Twelve Step Program distilled a single insight about the possibility of a spiritual transformation. In hindsight, it may be instructive that that the genesis of this insight emerged from the area of religion, rather than, as might be expected in the attempt to develop a treatment for a medically based disease, from the field of psychiatry. In a way that might be the opposite of today, at the outset *religion would take precedence over science.* Yet, in the same breadth, the intellectual climate was also ripe for extrapolating a therapeutic tool from religion that would have the same practical benefit as would any comparable medical or clinical approach.

Insofar as the Twelve Step Program is still practiced today as one form of treatment, two considerations immediately arise. First, if religion is to have this therapeutic benefit, then it is by virtue of its having a universal appeal, e.g., in its power *to speak* across cultural divides to address the human predicament as a whole, that the former can develop into the latter. Second, if this religious-therapeutic approach is to exhibit this universal appeal, it must be possible to glean from the mixture (or unique "*genealogy*" of its sources) the basis of this universality. My aim in this chapter, then, is to trace the crisscrossing of various intellectual paths that led to the Twelve Step Program. In this way, I will outline the cultural milieu *or interpretive horizon* from which the founder of that program first attempted to understand the problem of addiction. I will begin by identifying the journey along with Bill Wilson traveled in arriving at his "fateful" encounter with Carl Jung. I will then outline further the random confluence of events, the so-called "twist of fate," which brought together the key figures who pioneered the Twelve Step Program. Finally, I will lay bare the wider cultural-historical horizon out of which Jung's therapeutic practices, and religious precepts of the Oxford Group, converge to create a new brand of spirituality, which is non-denominational and has a global reach. As this interpretive horizon comes to the forefront, I will identify a gap that arises between the development of a therapeutic practice for treating addiction, and our understanding of it *as* a problem, namely, the existential ground of the patient's inherent resistance to seek treatment.

6.1 A Fortuitous Journey: Bill Wilson's Discovery of Carl Jung

During the period in which Heidegger came to prominence as the leading philosopher in Germany in the late 1920s, new tactics and methodologies for understanding religion, particularly, Christianity, were also being implemented, both in Europe and England. In outlining this confluence of these intellectual traditions, we can

[1] For an interesting discussion of the mythic import of necessity, and its hermeneutic implications for the development of self-understanding, see Risser 2012, pp. 8–9.

point to Heidegger's colleague at Marburg (Germany), Rudolf Bultmann (1884–1976). Bultmann not only influenced the former in the development of his hermeneutic method; but he (Bultmann) also incorporated it into his plan to "demythologize" religious texts, specifically the Bible, in order to extract the "core message" of Christianity.[2] In this way, Bultmann pioneered a method for restoring the contemporary relevance of Christianity, in order to show how spiritual insights can illuminate the human predicament (even the problem of addiction), without invoking the authority of any specific religious denomination. As we will see, in developing the first treatment program for addiction, Bill Wilson shared elements of this common intellectual and cultural vision.[3] The prototype for his plan for spirituality, however, would be imported simultaneously, not from Germany, but from Great Britain in the guise of the "Oxford Group" of the 1920s.

For those familiar with the saga of Bill Wilson's journey, the Oxford Group bears the name of an important catalyst both in his recovery from addiction and in his formulation of the basic principles of the Twelve Step Program. Because in academic circles the term "Oxford" can easily be linked to a brand of analytic philosophy, there is also room for confusion—not to mention the fact the "Oxford Group" is virtually unknown as a historical movement outside its influence on Bill Wilson. Yet, a key intellectual figure whom Bill Wilson actually came to know through correspondence, and whose prominence in the field of psychology would come to equal that of Heidegger in philosophy, was C. G. Jung. Indeed, Jung stands out both as an inspiration to Bill Wilson, and by showing how faith can function as a concrete "therapeutic" principle. By occupying a special place in the pantheon of intellectuals of his time, as Heidegger did within philosophy and Bultmann within theology, Jung provides an important linchpin in outlining the wider cultural-intellectual backdrop from which the Twelve Step Program evolved.

Bill Wilson was a pioneer within the social and intellectual fabric of the twentieth-century. And yet, for those who have not studied the problem of addiction, his name remains virtually unknown even by individuals who implement the Twelve Step Program he developed. Perhaps this discrepancy reflects the fact that the Twelve Step Program has been broadly applied to treat various forms of addiction, besides what it was initially designed to treat (e.g., addiction to alcohol). The Twelve Step Program thereby became separated from the community of fellowship, which Bill Wilson believed was the key to *implementing* this program, and which bears the name of the movement he pioneered: "Alcoholics Anonymous" (AA). Because there is a historical distance between the Twelve Step Program and AA, we need to examine the origin of the latter in order to appreciate the impact of the former. With an eye to Heidegger's hermeneutic phenomenology, we must consider the existential circumstances, as well as the chronology of events, which shaped Bill Wilson's journey, i.e., a path to self-discovery that culminated in his pioneering the Twelve Step Program. n surrounding AA as a group of misfits and outcasts.

[2] van Buren 1994, pp. 150, 152.

[3] See B. (Dick) 1992, p. 124.

There are many sources we can look to in order to identify the development and evolution of AA. And perhaps what proves most striking is the eclectic and diverse nature of these sources: none holds a monopoly on shaping the ideas of those who ultimately formulated the twelve steps of AA. And perhaps there is an important lesson to be learned from this fact. For if the problem of addiction to alcohol and other substances affects people world-wide, and if the program for confronting this problem can be viewed as viable, then it must originate from sources that in some way have *universal appeal*, that is, which can speak to people across all cultures and class distinctions. And while the manner in which the founders of AA come across these sources may appear to us in retrospect to be haphazard, it may also be the case that hindsight allows us to discern the crystallization of a distinctive "wisdom," as we say today, a "method in the madness."

The phrase "method in the madness" should not in the least be taken negatively, but rather points to a confluence of accidental factors that add up to good fortune. In seeking a solution to his own struggle with alcoholism, Bill Wilson wrote a letter to the prominent Swiss psychiatrist, C. G. Jung. As the story goes, Jung, frustrated with his own lack of success in treating alcoholic patients, replied to Bill Wilson that only a radical spiritual transformation or "conversion" could grant the individual the power and guidance to confront his/her alcoholism. The upshot of Jung's message may have been *ambiguous*, but perhaps intentionally so. He claimed that alcoholism was a "hopeless" illness, or, put another way, a disease "without hope." The illness was decidedly unique, because at some level of the person his/her own penchant toward self-destruction seemed to work against the possibility of any treatment, let alone recovery. Absent of any recipe for therapy or treatment, alcoholism appeared to be untreatable, both on the side of therapist and the would-be patient. While the addict experienced his/her plight as being "without hope," that *pathos* was simultaneously corroborated on the side of the therapist who viewed treatment as more or less futile, i.e., "hopeless."

For Jung, however, hopelessness did not simply translate into capitulation (and herein lies the ambiguity). For hope could also be reinterpreted as an acronym for a higher power or redemptive source, that is, for God in some indefinable way. To say that the alcoholic is "without hope" is to suggest that he/she is bereft of a higher spiritual connection, or what can ultimately restore a sense of hopefulness. Thus, the so-called despair of the addict could also be read symptomatically as the need for spiritual renewal and transformation. In retrospect, Jung's negative admission of his inability to help alcoholics (even Bill Wilson) harbors a positive testimonial that any catalyst for recovery begins with restoring a sense of hope, that is, through a connection to a higher power. For, as the example of the chronic alcoholic illustrates, the change in the behavior of one's drinking habits, in the face of a person's "human, all too human" foibles, may be difficult to reverse. Thus, the tendency of the alcoholic is to revert back to his/her old habits. This simple fact, which Jung realized through his lack of success in assisting Rowland Hazard, and which Bill Wilson knew through experience, led to the radical conclusion that only a "conversion" could yield the *possibility* of more lasting results. Underlying this conclusion is the insight that any change in behavior must be accompanied by a dramatic change in one's *way of thinking. The change in one's pattern of thinking is the individual's own willingness to practice the 12 step-principles.* While it may subsequently be evident that a conversion-experience provided the foundation of any program for the recovery of alcoholics, we might very well reflect on this unique historical source and its possible irony. Indeed, of the various twentieth-century luminaries who contributed to the collective wisdom of AA, Jung is as prominent of anyone. Along with Freud, Jung pioneered the development of modern psychoanalysis and psychotherapy.

Why would Bill Wilson specifically seek the counsel of a psychiatrist, albeit a celebrated practitioner as Jung? What does this historical fact reveal to us about one of the chief sources in the development and evolution of AA? As a layman, Bill Wilson, appeals to an "expert" about a problem the key to resolving would seemingly fall within the specific field of psychiatry/psychology. The expert, himself having had no success treating patients afflicted with this specific problem, redirects Bill Wilson away from what that field alone has to offer to insights gathered within an alternative domain of concerns about the possibility of a life-transforming experience, namely, the arena of faith. Far from involving a treatment that is rationally definable by the objective method of science, a solution seems instead to hinge on a uniquely personal experience to the individual alone and which becomes meaningful only within the "extra-rational" domain of faith.

Though we will return to consider this spiritual dimension, what does the exchange between Bill Wilson and Carl Jung tells us about the phenomenon of alcoholism? If alcoholism is a disease, then it is a peculiar one insofar as it does not fall into any simple pattern of treatment, and, indeed, is not "curable" in any conventional sense of the word. But more to the point, if alcoholism cannot be reduced merely to physical or even psychological symptoms, it may instead constitute a problem that pertains to the whole person. To the extent this is the case, then not only is alcoholism a stubborn problem; but, it may also be the case, conversely, that unraveling the problem of alcoholism can illuminate both the heights and depths of the human spirit.

In 1933, a year before Bill Wilson asked for his advice, Carl Jung published a collection of essays titled *Modern Man in Search of a Soul*. In the concluding essay of that volume, Jung addresses the topic "Psychotherapists or Clergy."[4] For centuries, individuals in search of answers to the basic quandaries of life, who have fallen prey to the greatest forms of spiritual suffering, have solicited the advice of priests and clergyman. These "emissaries" of God professed a wisdom that could provide insight and guidance to those who have long lost their ways in life. But at the beginning of the twentieth century, a secular alternative to pastoral counseling emerged in the guise of modern psychoanalysis and psychiatry. Spearheaded by (1) a tendency toward introspection and (2) a weakening of religion in general, and Christianity in particular, as a structure of belief or dogma within which the individual could find both guidance and security, psychiatry emerged as a new paradigm for addressing the problems of the self. Given this historical development, the individual turns, not

[4] Jung 1974, "Psychotherapists or Clergy," pp. 221–244.

a spokesman of the Church for spiritual direction, but to a specialist within the medical profession, the psychiatrist. As Jung emphasizes, psychiatry does not simply offer a secularized form of religion. Rather, psychiatry provides a secular analogue to an important role that religion has traditionally fulfilled, namely, a personal spiritual advisor that religious institutions have traditionally reserved to the "rabbi," the "pastor," the "priest," and even in its de-institutionalized form, the "guru."

While initially the tutelage of psychiatry and religion may be at odds, the increasing climate of secularization begins to blur the lines between the two. And yet the end of that process is not simply for one form of spiritual practice to overtake and replace the other. Rather, the fact that psychiatry can assume its place alongside religion only means that there is a common denominator between the two. For the "psyche" of psychiatry is just another word for "soul," that since the early days of Christianity has been the focus in the religious striving for atonement, resurrection, and salvation. We may speak of "psychopathology" to distinguish a specific kind of problem, which is open to medical diagnosis, e.g., obsessive-compulsive disorder. Similarly, we may allude to an individual immersed in a spiritual conflict as a forlorn or a "sick soul," suggesting instead what clergy describe as a "crisis of faith." In either case, there is a malady or problem afflicting the "self" that begs for some kind of intervention with the aim of alleviating the distress, that is, of initiating some process of transformation and healing.

In today's language, the word "healing" has primarily medical connotations; it refers to the process whereby a patient undergoes treatment in the hope of "curing" a disease. But the ancestry of the word "healing" also suggests an earlier meaning of making the person whole again, after a period of suffering imbalance and fragmentation. In the derivation of its root, then, "healing" exhibits a common ancestry with another word, albeit of a distinctly spiritual character, namely, "holy." With this more original sense of "healing" in mind, the practice of medicine need not be detached from treatments that address the well-being of the person as a whole. Such treatments may be multidimensional insofar as they address the physiological, psychological, and spiritual aspects of the person, not haphazardly, however, but in regard to the individual's search for unity and wholeness.[5] Thus, a therapy centered on the whole person (e.g., the ecstatic openness beyond the dualism of mental and physical), and a "principle of healing through meaning, is grounded on…the phenomenological insight of the meaning-giving-receiving-discovering-fulfilling activity of (life) of human consciousness."[6] As Jung remarks:

> Healing may be called a religious problem....
>
> Man is never helped in his suffering by what he thinks for himself, but only by revelations of wisdom greater than his own. It is that which lifts him out of distress.
>
> Today this eruption of destructive forces has already taken place, and man suffers from it in spirit. That is why patients force the psychotherapist into the role of a priest, and expect and demand of him that he shall free them from distress. This is why we psychotherapists must occupy ourselves with problems that strictly speaking, belong to the theologian. But

[5] See Frankl 1980, pp. 11–22.

[6] Kovacs 1982, p. 37.

we cannot leave these questions for theology to answer; the urgent, psychic needs of suffering people confront us with them day after day.[7]

The exchange between Bill Wilson and Carl Jung is not only a historically interesting step in the development of AA. Indeed, their correspondence casts light on a unique catalyst and turning point in the actual formulation of the principles of AA. Not only did Bill Wilson seek guidance from Jung, but Jung also outlined the cultural climate in which Christianity could be reinterpreted as a manual for personal spiritual growth and development. The ongoing secularization of religion, which I will return to address, contributes to this cultural climate. But the other side of the coin is that neither Bill Wilson–nor Jung for that matter–was a philosopher or/theologian. On the contrary, both held in common an extremely "practical" agenda. On the hand, Bill Wilson, a stockbroker by profession, was grasping for straws for some plan, on even inspiration, to address the vicissitudes of addiction to alcohol. On the other hand, Jung was interested in Christianity and other world religions, not as a dogma or system of beliefs, but insofar as the various religious traditions spoke to the universal character of the human predicament, the perennial drama of the struggle between suffering and salvation. He thereby saw in religion the constellation of many insights into humankind's spiritual predicament, which he sought in turn to *apply* with the end of eliciting an individual's personal transformation within the context of psychotherapeutic dialogue.

In retrospect, Bill Wilson and Carl Jung were drawn together by a common concern for spiritual awakening of the "self" *and* the applicability of the insights of Christianity toward this practical end. Wilson and Jung were pragmatists par excellence, even though neither could be called disciples of the American school of "pragmatism." In accord with Jung's appeal that a conversion was necessary to break the bond of addiction, Bill Wilson turned to one of the fathers of American pragmatism, William James, to find a detailed account of the phenomenon of religious conversion.[8] As a historical footnote, Jung, as well, carefully read James's pivotal work, *The Varieties of Religious Experience*.[9] But while neither fell under the "pragmatist" label, Wilson and Jung both epitomized the pragmatic spirit. In a more explicit way than James, however, both emphasized the therapeutic value of religion. But then the question arises: what form must religion take, whether it be Christianity or one of its variations, in order to be amended to facilitate the end of psychotherapy? And as the example of Jung already illustrates, only a universal form of Christianity, crystallized into its most elemental insights, and, just as importantly, stripped of its dogma and sectarian biases, will do.

A universalized Christianity may not proclaim the existence of God save as a greater power than the individual, on which the possibility of his/her spiritual transformation rests. Given his practical bent, Jung circumvented esoteric debates about the existence and reality of God. Within the context of his therapeutic practice, Jung

[7] Jung 1974, pp. 237, 240–241.

[8] See B. (Dick) 1992, p. 28. See Kurtz 1991; pp. 33, 182. Also see O'Connor 2016, pp. 157–158.

[9] See Jung 1958, p. 8.

considered the Divine only as a mystery that becomes manifest through the profound changes that the individual undergoes on the path of spiritual transformation. Accordingly, Jung sought the locus for this power, not in the ether above, but in the deepest recesses of the human psyche, in what he called the "archetypes of the collective unconscious."[10] To be sure, Jung's theory of human psyche was not central to the development of the principles of AA. Yet, when coupled with the practical emphasis of his therapy, Jung's effort to address the Divine as an immanent spiritual power, rather than as a transcendent deity clothed in dogma, provides a key example of the overview from which the founders of AA would extract the spiritual insights to forge its own basic principles.

Though AA refers to God, there is always the accompanying qualifier "as we understand him."[11] Bill Wilson and the founders of AA adopted a version of Christianity whose insight could be distilled in the most universal form possible. In this way, a reference to a greater power could be maintained without the risk of polarizing the members of AA into specific religious denominations and 2) inviting participation even among those who disavow any belief in a deity, who might be considered either "agnostic" or "atheistic" in the ordinary sense. Indeed, while the result of sectarian religion is to divide, one of the primary missions of AA is to overcome or avoid such divisions altogether. The goal of AA is to be inclusive, not exclusive. And this end is best served by distilling the central message of Christianity to its most basic insights, in order to convey the practical messages of spiritual change to the alcoholic without creating any further psychological or social barriers in the process.

6.2 A Fateful Turn in the Development of the Twelve Step Program

6.2.1 Therapeutic Origins

Though it was probably not viewed as such at the outset, the universalist version of Christianity from which Bill Wilson and the other founders developed the principles of AA, would spring from the Oxford movement. The connection between AA and the Oxford movement of the 1930s is well-known and well-documented. The philosophical and theological roots of the Oxford movement, which I will consider in the next subsection, are probably not as well known to the general public. But however we trace the evolution of AA, the link that cannot be underestimated lies in the influence that an alcoholic patient of Jung's had on Bill Wilson, both in shining a beacon in his personal search for sobriety and in supplying the germ for the spiritual program of AA. The patient's name was Rowland Hazard. Hazard was the

[10] See Jung 1965, p. 161.

[11] See B. (Dick) 1992, pp. 91, 100 ("Big Book").

intermediary, or, as we would say today, the "contact person" between Bill Wilson and Carl Jung. Moreover, at Jung's recommendation that only a "spiritual solution" could speak to his alcoholic problem, Rowland Hazard joined the Oxford Group, an arm of the Episcopal Calvary Church of New York. Hazard, through another member of the church he tried to help with his problem with alcohol (Ebby T.), not only relayed to Bill Wilson Jung's wisdom about the central importance of a religious conversion. But, spearheaded by the Oxford Group and his association with Ebby T., Hazard also would supply a framework within which to reap the fruits of Jung's suggestion.

In the figure of Rowland Hazard, then, we discover a link between Bill Wilson, Carl Jung, and the Oxford movement. In search of a cure for his alcoholism, Hazard had traveled to Zurich, Switzerland to undergo psychotherapy with Jung. When Jung realized that none of his therapeutic practices could help Hazard, he (Jung) strongly suggested that his (ex-)patient seek his own spiritual transformation through the support of a religious group. Though it is not known that Jung specifically suggested the Oxford movement, the end result is that Hazard's association with that group became the crucial step in initiating Bill Wilson to a new spiritual program. For Hazard became acquainted with the Oxford Group through the Rev. Sam Shoemaker, the minister of the Calvary Episcopal Church (New York), which Hazard joined after Jung pointed him in a new spiritual direction. Hazard's affiliation with Sam Shoemaker become especially important, insofar as Shoemaker had been a disciple of Frank Buchman, the individual credited with founding the Oxford Movement in England and a student of the eminent Professor Henry Drummond of Scotland.[12]

Yet while Hazard provides the key linchpin in the early founding of AA, there is a curious twist to the story of Rowland Hazard that makes his influence upon Bill Wilson appear fortuitous to say the least, and, the founding of AA hinge on a pivotal twist of fate. Initially, Rowland Hazard had sought the help of Jung's illustrious mentor and founder of modern psychoanalysis, Sigmund Freud. Hazard also solicited the assistance of another major figure in psychoanalysis, Alfred Adler. But when Freud, too ill to lend his services to help Hazard, and Adler as well was unavailable, Rowland Hazard solicited the advice of Carl Jung. The fact that Jung was available, and Freud and Adler, were not, became a fortuitous turn of events. For Freud, with his claim that sexuality was central to human behavior, and Adler, similarly, that the "drive to power" served a similar role, were neutral, if not downright hostile to religion. Carl Jung, on the other hand, believed that religion was crucial to understanding the path of self-development, and it is this religious emphasis that became the major point of contention precipitating his break with Freud, and Adler as well.

Along with this best known among intellectuals for his stark criticism of religion. Perhaps a single statement from his famous work, *The Future of an Illusion*, most aptly summarizes his vehement attack on religion: "Religion is the universal obsessional

---

[12] For an account of these historical connections, see B. (Dick), (1992), pp. 36–55.

neurosis of humankind."[13] In breaking with his mentor, however, Jung took a completely different position on religion. In his essay "Freud and Jung," Jung succinctly summarizes the differences that separate his approach to psychotherapy from that of both Freud and Adler:

> Both schools, to my way of thinking, deserve reproach for over-emphasizing the pathological aspect of life for interpreting man too exclusively in terms of his defects. A convincing example of this in Freud's case is his inability to understand religious experience, as is clearly shown in his book, *The Future of an Illusion*.[14]

Of course, we can only speculate on the influence of a hypothetical. It appears safe to extrapolate, however, that an expert like Freud who viewed religion as a mere "crutch" for the weak, indeed, as a form of psychopathology, could have been little help in suggesting to Rowland Hazard that a spiritual renewal might hold the keys to breaking the shackles of addiction. Where Jung could appreciate Hazard's initiative in seeking the "Christian Fellowship" through the Oxford Group, in all likelihood Freud would view the same as a symptom of an inability to cope with reality, that is, as a sign of psychological regression.

Fate, or something even more personal in the name of Providence, sometimes directs a person along the proper path. Such was the case with Rowland Hazard. Through his acquaintance with Jung, Hazard sought the fellowship and assistance of the Oxford Group. And this unlikely turn of fate then became one of the most important catalysts in transforming the Christian Fellowship of the Oxford Group into the network of AA. We would be amiss, however, to suggest that this network emerged overnight as strong and comprehensive. On the contrary, the historical facts suggest just the opposite: individuals on the brink of despair, including Rowland Hazard, were seeking any kind of wisdom and assistance to help give new direction to their lives. Indeed, the fraternity that would ultimately come to be called "AA" included as one of its founding, albeit fragile members, a "mutual friend" of Hazard and Bill Wilson, Ebby T. (Thacher).[15] After Hazard intervened in Ebby Thacher's behalf in order to prevent the latter from being committed due to his drunkenness (in 1934), Ebby T. soon after met with Bill Wilson to relay the insights that Hazard gleaned into the plight of alcoholism. Not only did Ebby T. transmit to Bill Wilson the lesson of Hazard's contact with Jung as well as the fraternity of the Oxford Group; in cooperation together the three men, bound by a commitment to overcome their addiction to alcohol, were also forming the fabric of a community that would culminate in the creation of AA. In the language that would become the foundation of AA, Rowland Hazard served as Ebby T.'s sponsor, who in turned "sponsored" Bill Wilson in his search for recovery.

While our primary focus lies in uncovering the origin of the ideas which translate into the principles of AA, we cannot underestimate the importance of the formation of the actual community of its members and of the key belief that motivated Bill

[13] Freud 1959, p. 55.

[14] Jung 1974, "Freud and Jung—Contrasts," pp. 115–124.

[15] Kurtz 1991, p. 9.

Wilson: the need to "pass on" the collective wisdom of the group to new individuals in needs. The spirit of the new maxim became: "You have to 'give it away to keep it.'"[16] Indeed, the ideas themselves would remain merely abstract and extraneous if it were not for the *resolve and commitment of those sought to put those ideas to practice in the form of the actual principles of AA*.

Through the collective wisdom of Jung and members of Oxford Group, along with the concerted effort of Rowland Hazard, Bill Wilson, and Ebby T. to apply those insights in the hope of breaking their own dependence on alcohol, a new era was dawning in the treatment of alcoholism. Where alcoholics once suffered in a vacuum of indifference and disdain, a new fellowship was emerging that could give assistance and guidance to those who previously had none. As history shows, "providence" had opened up a new path of hope for those previously lost in despair. Perhaps this historical moment of change can best be summarized in a few lines of the German poet, Friedrich Hölderlin, whom Jung quotes:

> Danger itself
>
> Fosters the rescuing power.[17]

As the simple description of one of Jung's foremost contemporaries, Martin Heidegger suggests, Hölderlin emerged as the "poet of all poets," because he juxtaposed a religious vision of hope with a Greek understanding of tragedy.[18] While Heidegger and Jung remain separated by presuppositions of their respective fields, both recognized the importance of the Greek *mythos* in providing a historical backdrop to understand the human predicament and develop a *logos*, or language to express the opposite possibility of "salvation."[19] In the context of Heidegger's phenomenology, however, Bill Wilson's saga testifies to an important "existential" fact: that whatever the presumed source of any personal salvation may be, whether supposedly "transcendent" or otherwise, the corresponding change in self-understanding and self-awareness still unfolds from the roots of our temporal sojourn, the journey of our being-in-the-world. Indeed, the fortuitous character of Bill Wilson's journey provides an illustration, if not of Jung's concept of "synchronicity," then certainly of Heidegger's concept of the "moment of vision" (*Augenblick*) or "original simultaneity," of that decisive juncture where future and past converge to throw open a horizon of possibilities and "clear" the way to act upon them in a creative way.[20]

Though in philosophy we tend to use the word "fate" hesitantly, the transforming of the circumstances of one's adversity into new opportunities serves as an existential example of such "fatefulness" or "destiny."[21] If there has even been an instance of negativity breeding its opposite, and a radically new "life-direction" emerging as

---

[16] See B. (Dick) 1979, p.124.

[17] Jung 1974, "The Spiritual Problem of Modern Man," p. 220.

[18] See Heidegger 2000a, p. 52.

[19] For an excellent discussion of parallels (yet differences) between Heidegger's and Jung's views of the self, see Capobianco 1993, 50–59.

[20] Heidegger 1985, p. 113.

[21] See Zimmerman 1986, pp. 135–145.

a result, then the legacy of Bill Wilson certainly qualifies as such. We thereby witness a unique case of what Heidegger calls the hermeneutic "circle," in which Bill Wilson recovered from his addiction by following a path, which in turn opened the way for developing a "program" that others could follow in their struggle both to understand and manage their addictions. By showing how Bill Wilson's saga unfolds within a wider circle of understanding, we bring to light the interpretive horizon within which to address the germ and development of the insight that gave birth to his Twelve Step Program. Let us expand this horizon by examining the cultural milieu out which one of its foremost building blocks emerges, the Oxford Group, and make explicit the *philosophical underpinnings* of its origin (which have largely been ignored).

6.2.2 Revisiting the Oxford Group

The Oxford Group includes both a religious outlook and an association of fellowship, as we have seen. We can trace its roots back to its founder Dr. Frank Buchman (1878–1961). More an evangelist than an academic, Buchman traveled extensively, meeting such spiritual leaders as Gandhi in India (1915).[22] Buchman's evangelical fervor eventually led him to Cambridge, England, where in 1921 he formed a small association of individuals who came to be called the "Oxford Group."[23] The seeds of the Oxford Group (also called the "Moral Rearmament Movement"), as we know it today, took root during Buchman's stay at Cambridge when he founded "A First Century Christian Fellowship" in the fall of 1922.[24] As history shows, the Oxford Group developed strongholds in the Northeastern regions of the United States, for example, through the leadership of the Rev. Samuel Shoemaker of the Calvary Episcopal Church of New York. Frank Buchman, who had been an advisor to the YMCA on the campus at Penn State University, brought the Oxford Group to Bill Wilson's hometown of Akron in January of 1933.[25] Buchman was invited to Akron, not at the request of Wilson himself, but of Harvey Firestone. Harvey Firestone solicited the help of various spiritual leaders, including Shoemaker, to help his son, Bud Firestone, confront his problem with alcoholism. As unlikely place as Akron, Ohio became the point of intersection between the leaders of the Oxford Group and a few laypersons committed to confronting a vexing problem.

It is important to emphasize that the Oxford Group had no explicit connection to Oxford University,[26] having as its primary "academic" tie Professor Henry

[22] B. (Dick) 1992, p. 77.

[23] B. (Dick) 1996, p. 85.

[24] B. (Dick) 1996, p. 85. Also see "Editor's Note" (#6) in Jung 1958, p. 184.

[25] B. (Dick) 1992, p. 7.

[26] Frank Buchman's Oxford Group bears no resemblance to the Oxford Philosophy that became popular in Great Britain during the 1920s and 1930s. As practiced by two of its representative figures, G. E. Moore and Bertrand Russell, the Oxford Philosophy clashed with Christianity and

Drummond of Edinburgh. Frank Buchman, however, gleaned insights from Drummond only through his writings, never having taken courses from the esteemed professor. Though most accounts credit Henry Drummond as the intellectual well-spring of the Oxford Group, little attention is given to the historical cultural background for the germ of the insights which later inspired Buchman and others. Situated in Great Britain, Drummond formulated his thinking at the crossroads where the traditional beliefs in God were challenged by the rise of Charles Darwin's materially based evolutionary theory. Although many viewed Darwin's theory of evolution as heresy, some of Drummond's contemporaries, particularly intellectuals like Thomas Huxley, embraced the vision of natural and human evolution.[27] Drummond was specifically intrigued by the possibility that there could be spiritual laws governing the universe that parallel the laws of nature and the progressive course of evolution, which exhibits a spiritual analogue through the unfolding of God's plan throughout history.[28]

Drummond's opponents were not so much atheists, but rather earlier "deists" like William Maker. The deists claimed that God was like a giant watchmaker who created the world, only to remain aloof from it and allow it to run according to mechanistic laws like a watch. In Drummond's view, this overly intellectual view of God remained too narrow, so long as it relied solely on rational proof and neglected the mysteries of grace to which the Bible gave witness. On the contrary, salvation and the love expressed through it was an indispensable part of the Divine reality. For Drummond, no account of our natural origins could stand on its own, without leaving open the possibility of a further exposé or narrative about humanity's perennial struggle for salvation, and God's participation in offering a solution. A passage that Drummond cites from a treatise titled "Paradoxical Philosophy" aptly summarizes his vision of the intersection between spirit and nature:

> What we require is no new Revelation, but simply an adequate conception of the true essence of Christianity. And I believe that, as time goes one, the work of the Holy Spirit will be continuously shown in the gradual insight that the human race will attain into the true essence of the Christian religion. I am thus of opinion that a standing miracle exists.... a direct and continued influence exerted by the supernatural on the natural.[29]

To employ a term prominent in nineteenth century Biblical scholarship, human history could only be understood in connection with the "history of salvation" (*Heilsgeschichte*).[30] Science and religion need not conflict. But they are different insofar as religion depends upon a language that postulates "absolutes," which cannot be directly proven either by logic or by an appeal to observable fact.

---

most expressions of faith. Indeed, the "Oxford Philosophy" dismissed any religious claims that could not be empirically verified. Russell in particular went so far as to argue that any symbolic and spiritual imagery, which could not be reduced to logical statements, was all but meaningless.

[27] Thomas Huxley, the grandfather of Aldous Huxley, occupied the same intellectual circle with Henry Drummond, Buchman's mentor.

[28] Drummond 1885, pp. 69n–70n.

[29] Drummond 1885, p. 60.

[30] Gnuse 1989, pp. 7–12.

Except for the accidental link to Bill Wilson through the Oxford Group, Drummond had no direct influence on Jung. Nevertheless, Drummond's interest in reconciling religion and science took a practical twist in Jung's attempt as a medical doctor and a psychiatrist, who maintained that the development of spirituality was central to any "therapeutic" practice. With his gnostic leanings, Jung opposed the "materialism" of the twentieth century that sought to reduce the person to a combination of instincts, drives, urges, and a conflation of chemical-physiological processes. As Jung states in questioning the "scientism" of his times:

> But people who are not above the general level of consciousness have not yet discovered that it is just as presumptuous and fantastic to assume that matter produces mind, that apes give rise to human beings, that from the harmonious interplay of the drive of hunger, love, and power Kant's *Critique of Pure Reason* should have emerged, and that all this could not possibly be other than it is.[31]

If "matter does not produce mind," then perhaps the affiliation of the human mind with a "higher power" may indicate the importance that "absolutes" play in the development of a spiritual program to tackle the pervasive problem of addiction.

The Oxford Group's influence on the development of the Twelve Step Program becomes most explicit through the "Four Absolutes." The Rev. Sam Shoemaker was a great proponent of them.[32] Most simply, these include: "absolute purity," "absolute honesty," "absolute unselfishness," and "absolute love." It is well documented, for example, that the founders of AA put great stock in Drummond's appeal to the famous passage in Corinthians: faith, hope and love, and the greatest of all is love. This Biblical quote becomes the source of inspiration for Drummond's book, *The Greatest Thing in the World*.[33] The reference to the absolute serves a dual role of (1) emphasizing the need for one's complete capitulation to a higher power and (2) emphasizing a corresponding *practice* of unrelenting rigor and candor, which culminates in what spiritual leaders from Zen Masters to M. Scott Peck call the "disciplined life."[34]

As we glean from Anne Smith, Dr. Bob Smith's (Bill Wilson's collaborator) wife, the Oxford Group's "Four Absolutes" formed the cornerstone of the practical guidelines of AA. "Bill Wilson claimed he had included the Four Absolutes in Steps Six and Seven, and in the use of the word God."[35] To cite these steps:

> 6. We are entirely ready to have God remove all these defects of character.
>
> 7. We humbly ask him to remove our shortcomings.[36]

And yet the importance that the Four Absolutes played in the early development of the spiritual program of AA became one of the *chief bones of contention that*

[31] Jung 1969, p. 341.

[32] B. (Dick) 1992, p. 32.

[33] B. (Dick) 1992, pp. 51–53.

[34] Peck 1979, pp. 78, 81–84.

[35] B. (Dick) 1992, p. 34

[36] B. (Dick) 1996, pp. 312–313.

threatened the marriage between the founders of AA and the Oxford Group. On the one hand, Bill Wilson emphasized that the Four Absolutes stood at the heart of the spiritual program. On the other hand, he quickly realized that the term "absolute" was a stumbling block for many alcoholics, who were intimidated by its sternness and lofty connotation.[37] The synonymous tie between the "absolute" and a requirement to be "perfect" also seemed be an obstacle for a spiritual program which needed to affirm the alcoholic's modest "steps" toward progress, no matter how tentative. Though the Four Absolutes remain vital to the development of AA, Bill Wilson's reservation about them illustrates an important obstacle he faced. Not only was discerning the content of the principles of AA crucial to its development, but equally as important was deciding upon the language in which they were expressed. Concerns about using the word "God," embraced by some and rejected by others, and the reluctance to allude to "Christ" at all, embody one of the most important quandaries that Bill Wilson faced.[38] Indeed, when words have long history, both within public consciousness and within an intellectual tradition–as religious terms certainly do–the connotations they carry can be as much impediments in facilitating an individual's recovery as aids.

With this caveat in mind, what do the Four Absolutes tell us about the development of the framework of the plan for recovery of AA? In the "four absolutes" we find one of the best examples of the spiritual program of AA where religion intersects with psychology, theory with practice. As such, they transpose the need to cultivate a relationship to a "higher power" into practical guidelines for initiating a "new life." Within the tradition of the Oxford Group,[39] Henry Drummond, and his predecessors, used the term "absolute" as a synonym, or even a replacement for, the word "God" (a precedent that Bill Wilson and Bob Smith implicitly followed in alluding, in Step Two, to "a power greater than ourselves").[40] In the nineteenth century intellectual tradition that prefigures Drummond's, although geographically located on the European continent rather than Great Britain, German thinkers like G. W. F. Hegel and Friedrich Wilhelm Schelling coined the term "Absolute" to define an idealist vision in which God stood for a unity encompassing all reality, and, conversely, nature and human history provided the spatial-temporal nexus for revealing the Divine.[41] The tradition of German idealism thereby set the precedent for distinguishing the reality of the Divine, as well as the dynamics of its influence upon us, without necessarily restricting the experience of God to any one of the world religions. Indeed, at the close of his epic work, Hegel employs the Christian imagery of crucifixion and resurrection, the "*Calvary* of the Absolute Spirit," to describe the pinnacle of God's self-revelation within the crucible of history.[42] With

[37] Kurtz 1979, pp. 25, 51.

[38] W. (Bill) 1953, p. 34. As the Third Step illustrates, we surrender to "the care of God *as we understand him*."

[39] B. (Dick) 1992, pp. 22–23.

[40] B. (Dick) 1992, p. 34.

[41] See Hegel 1974, pp. 124–131. See Schelling 1994, pp. 225, 239.

[42] See Hegel 1976, p. 493 [emphasis my own].

his own twist on the German brand of "idealism," Drummond and his follower, Frank Buchman, adopted the Hegelian precedent of referring to God in such terms as a "higher power" or "absolute."

But once again, the chief heir to the German tradition, as far as his influence on Bill Wilson goes, was C. G. Jung. While much of Drummond's thinking, prior to Buchman, was pitched to professional theologians and academics, Jung pondered less the speculative mysteries of the Divine (although he was familiar with the writings of Kant, Hegel, and Nietzsche); Jung instead developed a *therapeutic language that translated the "Absolute" into a transpersonal source for the unity and wholeness of the human self.* Accordingly, he emphasized the dynamics of personal change that originate through the individual's seeking a relationship to a higher source of unity or wholeness. For Jung, whether we can "prove" the existence of this source through rational argument, or even whether we refer to it as "God," is completely beside the point. What matters is that a relationship to something "transpersonal" is necessary in order that the individual person can achieve a measure of wholeness for his/her own self.

Because of his practical emphasis, and his refusal to commit to any one school of thinking or even a religious outlook, Jung draws from diverse sources of both Western and Eastern traditions: including Buddhism, Taoism, central philosopher figures from Plato, to Kant, to German idealism and Nietzsche, along with his own arcane interests in gnosticism. Jung credits his eclecticism as the intellectual light that directed his brand of therapeutic practice.[43] And perhaps it is this breadth of this religious vision that allowed Jung to glean from various sources insights that he could appropriate within the practical context of therapy. But most of all the reverence for this collective wisdom enabled Jung to acknowledge his humility as a therapist before the transpersonal powers of healing, which alone could give an enduring solution to an individual's spiritual quandary. As Jung emphasized, spiritual healing and recovery involves a lengthy journey, rather than a destination:

> The paradoxical remark of Thales that the rust alone gives the coin its true value is a kind of alchemical quip, which at bottom only says that there is no light without shadow and no psychic wholeness without imperfection. To round out, life calls not for perfection but for completeness; and for this the "thorn in the flesh" is needed, the suffering of defects without which there is no progress and no ascent.[44]

Perhaps it was the humility of this insight, as much as his own sense of helplessness, which prompted Jung to threw up his hands in the face of Rowland Hazard's plight, and acknowledge that all the psychotherapy in the world could not help Hazard conquer his problem with alcoholism, that is, without the aid of a religious conversion.

[43] Jung 1965, pp. 101–104. Although reluctant to embrace any philosophical school, Jung recalls the importance of his students days when he grappled with the writings of Plato and Kant, and, in fact, questions Freud for never having studied Friedrich Nietzsche and yet developing a theory of human nature that challenges the latter's insights. "Freud himself had told me that he had never read Nietzsche...." (p. 153).

[44] Jung 1971, p. 406.

Analogously, whatever insights Bill Wilson incorporated from the Oxford Group, they prove valuable only when recast in the light of his own personal transformation. The more we delve into the breadth of sources from which Bill Wilson evolved the spiritual program of AA, the more we can appreciate the brilliance and ingenuity of the solution it offered alcoholics. Yet the converse is equally as important: without his initial spiritual illumination or "white light experience,"[45] all the exhortations of preachers and philosophers would not have been sufficient to allow Bill Wilson to make the breakthrough that he did.

6.3 Revisiting the Roots of Faith

Historically speaking, there is no direct influence, one way or another between Heidegger's thinking and the religious principles of the Oxford Group. Yet, as we will see, Heidegger, through his association with Rudolf Bultmann, and his effort to distill the "core message" (Kerygma) of Christianity,[46] may provide important clues for grasping the wider historical origin and philosophical backdrop from which the insights of the Oxford Group took root and flourished. At the outset of his career, Heidegger was strongly influenced by Catholicism, having begun his theological and philosophical studies at a Jesuit Seminary. (1909).[47] Having left the seminary soon after entering it (due to illness), he pursued a doctorate in philosophy, only to find equal inspiration through such key Protestant leaders and thinkers as Martin Luther and Søren Kierkegaard. The latter's characterization of faith as a "leap" continues to shape Christian discourse (and practice) today. Heidegger coined the term "primal Christianity" to describe the self's return to the lived-experience of faith, apart from the authority of any institution or dogma.[48] This path of self-discovery exemplifies the temporal dynamic by which authentic or owned existence becomes possible, that is, through the enactment of the decisive "moment" (*Augenblick*) in which the individual discloses "who" he/she truly is. While not his primary mission, Heidegger, along with his colleague, Rudolf Bultmann, marked the possibility of an ecumenical change, i.e., in how orthodox religion, specifically Christianity, could be both understood and practiced. The latter side of this development, or the change in how the individual's practices his/her faith, would become the leitmotif for a small gathering of individuals in Scotland and England who in the 1920s united to form the Oxford Group.

The idea that the individual's potential for transformation provides the lens through which to view the elemental truths of Christianity, and not the other way around, indicated a completely new way for approaching religion. While Heidegger's

---

[45] B. (Dick). 1992, p. 58.

[46] Bultmann 1961, "New Testament and Mythology," pp. 15–16. Of special interest is Bultmann's reference to Hans Jonas' work on gnosticism.

[47] See Schalow 2001, pp. 25–26.

[48] See van Buren 1994, p. 317.

innovations had their first wave of impact within philosophical circles, the influence of Bultmann's theology in Germany coincided with the emergence of the Oxford Group (even though there was no direct channel of exchange). For Bultmann pioneered his hermeneutical method (parallel to Heidegger's), which opened the way to reinterpret the scripture, e.g., the Bible, in terms of the priority of the "word of God." The "word," however, was not necessarily meant "literally," e.g., as the direct invocation of a transcendent being; nor could the "word" be reduced to any authoritative claim, e.g., dogma or liturgy. On the contrary, the question of the unique dynamics of the word itself, such that it can serve as the "emissary" of a divine message: i.e., the word (e.g., as "*logos*") that functions in a double sense as "intermediary," as the singularity of incarnation (e.g., of God in the figure of Christ) and the gathering together of Christ's followers (e.g., in a "community"). In this way, the transcription of the message is *balanced by its reciprocal appropriation*. And, since the need for this appropriation is always mandated in each historical era, the word must also be "renewed" in the language unique to a culture, even if as Bultmann realizes, this entails putting aside the "otherwordly" imagery which applied in another culture two thousand years ago.

In coining the term "demythologizing" to describe his unique approach, Bultmann also gave rise to the possibility of misunderstanding. For he does not claim that "myth" is not in some way important to encoding the language to express the Divine message. Rather, Bultmann argues that confusion arises when the icons and imagery of myth are posited as literal truths in opposition to what our culture holds today according to the laws of natural science.[49] To avoid this clash of science and religion, Bultmann suggested that myths enshrouded a unique genre of truth, or that of "faith," within a cryptic language, which had to be deciphered, as it were, in part by extracting the arcane, supernatural motifs in order to circumvent the aforementioned opposition. This is the "negative" side of demythologizing Christianity. The "positive" side involves allows the "meaning" of the "word," of divine scripture, to re-emerge in its contemporary relevance, in recovering the message of faith, the "Kerygma," in terms of its relevance for today. Thus faith can be "reborn," even in a secular world whose scientific outlook rejects "supernatural" explanations.

If the "word of God" is to assume this priority, then its dynamics must hold the key to unlocking the mysteries of the Kerygma, including the trinity. The singularity of the incarnation of the divine in the human, the preservation of the unity of the divine throughout this sacrifice, requires another permutation. For the word as an emissary of the divine, and of its "expression" in human form, also implies radiance the spirit in each individual who together form a community of believers. The analogical appearance of the divine in each individual, the "indwelling" of the spirit through the fellowship of the community, constitutes the dynamic of the "holy spirit," the third member of the trinity. This analogy allows for the infusion of the "divine" spark in the human, "Christ in me," and, reciprocally the transformed way of dwelling through one's proximity to God, "I in Christ."[50] However finely we may

---

[49] See Bultmann 1959, pp. 5–17.

[50] van Buren 1994, pp. 157–201. Also see van Buren 1992, pp. 159–162.

wish to parse these theological distinctions, the one dimension which stands out, if only in its name, as the key to the Oxford *Group*, is the role of Christian "fellowship."

The Oxford Group takes shape as a return to a "First Century Christian Fellowship." The formation of such a community was based on a shared faith, that is, in heeding the basic precept that to respond to the word of God is to spread its "message," the "good news" (i.e., the gospel). Herein lies the birth of a so-called "evangelical" spirit, in which its members, in heeding the word, as commissioned to do God's work, are also prepared to serve as "missionaries" in spreading the gospel. Yet the basic difference between this prototypical form of evangelicalism, and what goes under the rubric of "fundamentalism" today, is the former's commitment on the widest circumference on faith. For the onus lies on the individual's self-understanding of what his/her faith stands for, rather than "obedience," or conformity to dogma, or the blind acceptance of authority. As a result, the *practice* of that faith translates into the broadest, non-denominational, i.e., universalistic form of Christianity. This universalistic form of Christianity allows for, although does not guarantee, the fostering of tolerance as much for those who forsake belief, even those most "down and out," as those who are already predisposed to believe.

Within the purview of the Oxford Group, Bill Wilson would find not only the germ for development the "fellowship" of those struggling with their addictions, he would also find in the vision of a universalist Christianity the key to recasting the traditional concept of God. Indeed, whether by accident or design, the Oxford Group's formula of a generic Christianity squared perfectly with the development of plan for treating addiction. For that "philosophy" gives room for each person to conceive of the Divine, without imposing a form of "orthodoxy," creed, or dogma, which most addicts vehemently resisted. Indeed, the idea of universal Christianity solved the dilemma of how to ignite in those struggling with their addictions a sense of spirituality without, however, "forcing religion down their throats." Thus, the precedent was set for conceiving of God, as became the custom within the language of AA, as a "higher power." The first of the "Twelve Steps" declares the need, for those who "have become powerless" in their addictions, to turn their lives over "to God, as we conceive him." That is, the language leaves it open for the individual to view God as he/she sees fit, even if defined by such secular concepts as (the pursuit of) "sobriety" or even "mental health." The vision of the divine is transformed from that of orthodoxy to an "operational," "therapeutic" concept. Luther's thinking in particular,[51] through his notion of "grace," perhaps best exemplifies the spirit of this governing principle of the Twelve Steps: namely, that divine healing is a "gift" that goes hand and hand with the individual's renouncing his/her willfulness. Does this principle merely suggest a form of mysticism, or can we outline the dynamics of this process of self-transformation and healing?[52]

---

[51] See McGrath 1985, pp. 55–64.

[52] In its historical genesis, the "mind over matter" principle emerges from a religious group called "gnosticism." The term "gnosticism" comes from the Greek word, "to know," or to have some special knowledge of or access to, a truth that is otherwise far removed, difficult to comprehend,

If Bill Wilson tailored his own program of treatment to include a spiritual thrust, Jung's psychoanalytic approach provides an interesting parallel. For Jung proceeded in the opposite direction to show how his own therapeutic practices, including his analysis of the archetypes of the collective unconscious and dreams, could provide a window into the spiritual realm. Yet, in a way that resonates both with the Oxford Group and Bill Wilson's Twelve Step Program, Jung cautioned against inferring the "reality" of God or extrapolating theological claims from specific psychic phenomena. In this regard, Bill Wilson and Carl Jung traversed different paths in reaching the crossroads where psychology and religion, therapy and spirituality, intersect. In his letter to Bill Wilson, Jung emphasizes that the term "spirits," which is often used as a synonym for alcohol, derives from the Latin word, "*spiritus*."[53] Jung points to the interface between psychology and religion, insofar as the treatment of an apparent "personality" disorder hinges on understanding its spiritual origin. He suggests that, contrary to the stereotype to the "bum" of on the street corner, the addict may not be someone who has sunk to the lowest level of humanity. Rather, the addict may instead be someone who has become a victim of his/her own excessive aspirations, which, while aiming toward a higher, spiritual fulfillment, have become misdirected and diverted into a lesser substitute. In his letter to Bill Wilson, Jung describes his assessment of Rowland Hazard's condition:

> His craving for alcohol was the equivalent on a low level of spiritual thirst of our being for wholeness, expressed in medieval language: the union with God.
>
> The only right and legitimate way to such an experience is hat it happens to you in reality, and it can only happen to your when you walk on a path which leads you to a higher understanding. You might be led to that goal by an act of grace or through a personal and honest contact with friends or through a high education of mind beyond the confines of mere rationalism.
>
> You see, alcohol in Latin is *spiritus* and you use the same word for the highest religious experience as well as for the most depraving poison. The helpful formula there is: *spiritus contra spiritum*.[54]

When viewed within this wider historical context, the legacy of Bill Wilson parallels that of Jung: if spirituality is necessary for personal transformation, so the path of recovery from addiction serves as a testament to our higher natures, that is, to the spirit.

More recently, as we have seen, Bruce Wilshire develops a similar insight in suggesting that "ecstasy deprivation" lies at the roots of modern addiction. He argues that individuals in modern society individuals turn to artificial substitutes in seeking a commensurate experience of exhilaration or "ecstasy," including a drug that, perhaps not coincidentally, bears that name. The search for a "high" is indeed ecstatic,

and even mysterious. As Hans Jonas states: "In the gnostic context, however, 'knowledge' has an emphatically religious or supranatural meaning and refers to object which we nowadays should calls those of faith rather than that of reason." Jonas 2001, p. 34.

[53] Segaller and Berger 1989, p. 18.

[54] Quoted in Segaller and Berger 1989, p. 181.

and whether in ancient or modern society, there are religious and ritualistic practices that emerge as socially legitimate avenues to achieve a sense of the "ultimate"[55] or seek a "higher," transcendent level of meaning.[56] Our phenomenological approach, however, provides a further context to understand the meaning of such "ecstasy," and to point to establish a wider philosophical as well as a religious, ritualistic basis for the use of that term. The "ecstatic" entails not only a transformative experience, as even the orgasmic consummation of sexuality might, but also a self-transcendence enacted within the wider compass of truth as openness. The individual experiences self-transcendence when his/his life-trajectory becomes, as Kovacs emphasizes, "horizon-opening."[57]

Thus, for Heidegger, the *ecstatic* moment strips the façade of the 'ego's' narcissistic desires, in such a way as to recover, nevertheless, the self in a new birth of being forthcoming, honest, and assuming "ownership."[58] The "ecstatic" marks the apex of the heightening and disclosing of the self's potentiality, albeit with a reciprocal awakening of individual's commitment to become responsible or "answerable" (*Antwortlichkeit*) for who he/she *is*. In his lecture course from the Summer Semester of 1936, Heidegger provides a clue to how we can reformulate the problem of addiction by pointing to a deeper origin in which the potential for healing and illness are intertwined. If we consider the locus of human desire as staging a conflict between the human aspiration (to achieve a higher unity) and the denigration thereof into a fragmented condition of egotistic craving, then addiction is an abysmal longing for fulfillment that simultaneously masks and mirrors the self's connection to a greater wholeness. Insofar as "addiction" (*Sucht*) is a "sickness" whereby the self errantly mistakes gratification for a higher yearning, then *healing begins by reinstating a humility whereby a new life-trajectory* (of aspiration) *can emerge from its opposite*.[59] By employing Schelling's dialectic of strife (*Streit*) and opposition, Heidegger imports a religious, metaphysical imagery of the self's engagement in a cosmic struggle of separating from and reuniting with the Divine—the ontological ground on which a theory of the "four absolutes" can be based.

The phenomenological attempt to understand the broader meaning of the "ecstatic" as the impetus toward spiritual aspiration and transformation provides an important bridge in bringing the discussion of this chapter to a close. The "ecstatic" becomes an important referent point and sounding board to understand the path of

[55] See Tillich 1957, pp. 1–5. Tillich defines faith as our "ultimate concern."

[56] As Kovacs emphasizes with reference to Frankl's thought, several "therapeutic techniques are based on the human capacity for self-transcendence and self-detachment." Kovacs 1990, p. 242. Also see Wilshire 2003, pp. 7–11.

[57] Kovacs 1986, p. 200.

[58] Here I am using the term in an "existential" sense referring to a development of "falling," which contracts the space of openness into the 'I" as the locus of the exclusivity of the self-concern—rather than in a psychoanalytic sense of "regressive identity"). As Heidegger points out, the more Dasein' s speaks in terms of "I," the less it is a self, 1977/1962, pp. In "A Philosophical Critique of Narcissism," Eugene T. Gendlin reinforces this point in emphasizing that the body includes its own capacity for expression that interfaces with the world. Levin (ed.), 1987: 251–304 (esp. 302–303).

[59] Heidegger 1985, pp. 124–125.

recovery from the fetish-like attachment of addiction, that is, as the temporally enacted process of self-discovery. We need to identify another dimension of this temporal enacted process, however, to make between the dynamic of "grace" and the Heideggerian claim of ownedness or authenticity. For Heidegger, owned or authentic existence does not arise from an exercise of will, but instead, requires a prior awakening to the openness of possibility which can inform any act of choice. This self-awakening involves relinquishing the *hubris* of having complete mastery over one's existence, in which a way as to mirror the Christian message of salvation or grace that "to win oneself one must first lost oneself." For Heidegger, ownedness is the "winning back" of individuality precisely by "losing" the façade of one's narcissistic desires. In this transformative moment, the self rediscovers its uniqueness on a higher plane (which, from the standpoint of "recovery," implies a "first step" of self-humbling, humility, and thereby receptivity to "grace"). As Heidegger recognized early on through his study of St. Paul's Epistles, the self can be freed to receive the abundance of life's possibilities only through a parallel concession of affirming its finitude.[60] Michael Zimmerman amplifies this insight with the following remark: "Authenticity occurs when *Dasein* lets temporality generate itself authentically, just as for the Christian regeneration involves a transformation of time (*kairos*) with which the individual is summoned to cooperate."[61]

The attempt to explore the *existential* underpinnings of a therapeutic and spiritual process of recovery marks a key transition in the attempt to understanding the *problem* of addiction in phenomenological terms. The implicit "gnostic" turn that the first plan for treating addiction raises a concern about the concrete fact that the addicted self experience as falling, as facticaly dispersed (*Zerstreutheit*) in its thrownness,[62] and through its *embodiment* vulnerable to fetish-like attachment of its desires (including dependency upon the use of physical substances such as drugs or alcohol). Phenomenological speaking, the body is still the locus of addiction, even if we understanding embodiment in concrete terms as the trajectory of our entrance into and openness to the world.[63] The question still remains as to how such ecstatic transformation clears the way for *unfolding a new life-trajectory* as the catalyst for healing and recovery.

6.4 Conclusion

By providing an intellectual genealogy of the development of the Twelve Step Program, I have uncovered the cultural milieu or *interpretive horizon* from which the founder of that program first attempted to understand the problem of addiction. The crisscrossing of these intellectual paths points to the possibility of recovering

[60] Schalow 2003, pp. 23–24.

[61] Zimmerman 1986, p. 194.

[62] Heidegger 1978/1984, pp. 172–173; tr. 137.

[63] See Schalow 2006, pp. 20–27.

the primal roots of Christianity, a task which Heidegger and Bultmann first initiated in Germany in the 1920s. Almost simultaneously, the Oxford Group in Scotland embarked along a parallel path, which carved its way in a great arc leading across the Atlantic vis-à-vis Bill Wilson's founding of AA (in Akron, Ohio in 1934) and then back to the European continent through his correspondence with the Swiss psychiatrist, C. G. Jung. This historical scenario provides an example of the "circular" movement of interpretation, the primary venue of hermeneutics, which unfolds in a dynamical fashion and does not presuppose any direct "causal" influence among the various thinkers. Yet, an unasked question still remains.

One of the hallmarks of phenomenology is that it considers indirect, even negative evidence, as implicating the importance of what is *not* said, expressed or explicitly done, in terms of unveiling the multifaceted character of addiction. While the intellectual journey leading to the first treatment program for addiction was a fortuitous one, an intriguing question still remains: what does Jung's reluctance to treat addicts, and, conversely, the fact that a therapy founded on religious precepts, rather than psychoanalytic ones, reveal about the problem of addiction, which was may have been disguised to the founders of the "Twelve Step Program?" Indeed, from a psychoanalytic perspective, addiction appears as a spiritual, psychological and physiological malady, insofar as the degree of the addict's dependency erects an unusual stubborn barrier in the application and receiving of any possible treatment. Is it psychological inertia, physiological/sensual gratification, spiritual weakness, or even other factors, including the enabling mechanisms of family, friends, and the institutions of society?

In asking these questions, we seek to cast a critical light on the founding of the first treatment for addiction, not in order to discount its importance, but, on the contrary, to uncover gaps in the development of the intellectual tradition from which it sprang. Is there another side of the therapeutic practice that is still to be uncovered, which speaks to the darkest corridors of the addict's penchant for self-deception and may amplify our understanding of addiction *as* a problem? As we will discover, the therapeutic practice employs untested precepts, e.g., testimonials of self-admission, honesty, which "work" on the basis of a religious orientation, but nevertheless are articulated in an intellectual and a philosophical vacuum. In this regard, our account of the genealogy of sources of that tradition also becomes a philosophical excavation or "archaeology" of sorts.

References

Bill, W. 1953. *Twelve steps and twelve traditions*. New York: AA Publishing.

Bultmann, Rudolf. 1959. *Jesus Christ & mythology*. New York: Scribners.

———. 1961. New testament and mythology. In *Kerygma and myth: A theological debate*, ed. Hans Werner Bartsch, trans. Reginald H. Fuller. New York: Harper & Row.

Capobianco, Richard. 1993. Heidegger and Jung: Dwelling near the source. *Review of Existential Psychology & Psychiatry* 21 (1): 50–59.

Dick, B. 1992. *Anne Smith's journal (1933–1939)*. Maui: Paradise Research Publications.

———. 1998. *The oxford group & alcoholic anonymous*. Maui: Paradise Research Publications.
Drummond, Henry. 1885. *Natural law in the spiritual world*. London: Hodden & Stoughton.
Frankl, Viktor. 1980. *The doctor & the soul: From psychotherapy to logotherapy*.
Freud, Sigmund. 1959. *The future of an illusion*, trans. James Strachey. New York: W. W. Norton.
Gendlin, Eugene T. 1987. A philosophical critique of narcissism. In *Pathologies of the modern self*, ed. David Michael Levin. New York: New York University.
Gnuse, Robert. 1989. *Heilsgeschichte as a model for biblical theology*. Lanham: University Press of America.
Hegel, G W. F. 1974. *On art, religion, philosophy: Introductory lectures to the realm of absolute spirit*, trans. J. Glenn Gray. New York: Harper & Row.
Hegel, G. W. F. 1976. *The phenomenology of spirit*, trans. A. V. Miller. Oxford: Oxford University Press.
Heidegger, Martin. 1978. *Metaphysische Anfansgründe der Logik im Ausgang von Leibniz*, GA 26. Frankfurt am Main: Vittorio Klostermann.
———. 1984. *The metaphysical foundations of logic*, trans. Michael Heim. Bloomington: Indiana University Press.
———. 1985. *Schelling's treatise on the essence of human freedom*. Athens: Ohio University Press.
———. 2000a. *Elucidations of Hölderlin's poetry*, trans. Keith Hoeller. Amherst: Humanities Press.
Jonas, Hans. 2001. *The gnostic religion*. Boston: Beacon Press.
Jung, C. G. 1958. *Psychology and religion: West and east*, trans. R. F. C. Hull. Princeton: Princeton University Press.
———. 1965. *Memories, dreams, reflections*. New York: Random House.
———. 1969. *The structure and dynamics of the psyche*, trans. R. F. C. Hull. Princeton: Princeton University Press.
———. 1971. Individual dream symbolism in relation to alchemy, trans. R. F. C. Hull. In *The portable Jung*, ed. Joseph Campbell, 323–455.
———. 1974. *Modern man in search of a soul*, trans. W. S. Dell and Cary F. Baynes. Harcourt, Brace, Jovanovich.
Kierkegaard, Søren. 1962. *Philosophical fragments*, trans. David Swenson. Princeton: Princeton University Press.
Kovacs, George. 1982. Phenomeonology and logotherapy. *Analytica Frankliana: The Proceedings of the First Congress of Logotheraphy*, ed. Sandra Wawrytko, 33–45.
———. 1986. Phenomenology of work and self-transcendence. *Journal of Value Inquiry* 20: 195–207.
———. 1990. Way to God as absolute transcendence. *American Catholic Philosophical Quarterly* 64: 240–249.
Kurtz, Ernest. 1991. *Not-God. A history of alcoholic anonymous*. Center City: Hazelden.
McGrath, A. 1985. *Luther's theology of the cross*. New York: Blackwell.
O'Connor, Peg. 2016. *Life on the rocks: Finding meaning in addiction and recovery*. Las Vegas: Central Recovery Press.
Peck, M. Scott. 1979. *The road less traveled*. New York: Simon & Schuster.
Risser, James. 2012. *The life of understanding: A contemporary hermeneutics*. Bloomington: Indiana University Press.
Schalow, Frank. 2001. *Heidegger and the quest for the sacred: From thought to the sanctuary of faith*. Dordrecht: Kluwer Academic Publishers.
———. 2003. Heidegger and the question of human addiction: A phenomenological approach. *Southwest Philosophy Review* 19 (2): 18–30.
———. 2006. *The incarnality of being: The earth, animals, and the body in Heidegger's thought*. Albany: SUNY Press.
Schelling, F.W. J. 1994. *Idealism and the endgame of theory: Three essays*, ed. and trans. Thomas Pfau. Albany: SUNY Press.

Seageller, Stephen, and M. Berger. 1989. *The wisdom of the dream*. Boston: Shambhala Press.
Tillich, Paul. 1957. *Dynamics of faith*. New York: Harper & Row.
van Buren, John. 1992. Martin Heidegger, Martin Luther. In *Reading Heidegger from the start: Essays in his earliest thought*, ed. T. Kisiel and J. van Buren, 159–172. Albany: SUNY Press.
———. 1994. *The young Heidegger: Rumor of the hidden king*. Bloomington: Indiana University Press.
Wilshire, Bruce. 2003. *Wild hunger: The primal roots of modern addiction*. Lanham: Rowman & Litttlefield.
Zimmerman, Michael E. 1986. *Eclipse of the self*. Athens, OH: Ohio University Press. (revised).

Chapter 7
In Search of a New Discourse: Resetting Priorities

As mentioned in the "Introduction," philosophy stands apart from other disciplines in its focus on the obvious, in its endeavor to bring into question what otherwise appears self-evident. What from the vantage point of psychology might appear to be inconsequential may from a philosophical perspective hold the key to a completely new understanding. Such is the case in emphasizing what is easily overlooked concerning the formulation and use of the "Twelve Steps," namely, their inherently linguistic character or their underlying dependence on a distinctive *logos*. That is, the Twelve Steps are a litany of statements, principles, and pronouncements. But more than that their very expression or utterance implies an activity to be performed (by the speaker), in such a way as to suggest a future precedent for changing (and thereby reversing) a habituated, past course of action. These twelve pronouncements, then, are a primary example of how language can open itself as a "performative" domain, the power of which is transmitted within the wider compass of one's worldly existence and thereby in concert with the flow of the self's *temporalizing* (e.g., the reversing of past habits by projecting a future possibility).[1]

As such, the first program for treating addiction implements a therapeutic process that is different from (although it can be supplemented by) traditional one-on-one "talk-therapies," as epitomized, for example, in the "private" relationship between clinician and patient. Even in the original AA program of group-fellowship and sponsor, the therapeutic process explicitly unfolds along the worldly axis of our being-with others (*mit-sein*). Hence, a key message of the twelve statements is to make amends with others as a catalyst on the way to recovery, and, implicitly, to avoid previous contacts and contexts that have previously proved to be detrimental such as bars, casinos, and their clients. Thus, while the Twelve Step Program may parallel talk-therapy, its unique *logos* implies a self-initiated, self-directed activity that radiates outward from each individual's potential for transformation into the

[1] For a further discussion of the temporal origin of language, and its power to express the self's existence through "formally indicative" examples, see Hatab 2016, pp. 275–277.

F. Schalow, *Toward a Phenomenology of Addiction: Embodiment, Technology, Transcendence*, Contributions To Phenomenology 93,
DOI 10.1007/978-3-319-66942-7_7

wider expanse of the world itself (including one's day-to-day endeavors, pursuits, family and friends).

As we begin to unravel the problem of addiction in all its complexity, our understanding of it also continues to evolve. In this chapter, I will first crystallize a richer understanding of this phenomenon, taking the initial steps toward re-defining addiction as it implies a corresponding awareness of the larger predicament of life. As such, the phenomenon of addiction is multifaceted. "'Where danger arises, also grows the saving power,'" distinguishes the words of the poet, as quoted by philosopher and psychologist alike. For the narrowing of the channels that lead an individual down the path of becoming addicted, in reverse, originate within a wider expanse of openness in which the power to choose can be exercised. Wherever freedom arises, the possibility of responsibility also emerges. Given this new orientation, I will, secondly, revisit a difficult question of the role that "responsibility" plays both in averting addictive behavior and in allowing an individual, once having crossed the threshold of addiction, to regain a sense of self-empowerment and self-direction. Toward the close of this chapter, I will show, thirdly, how our attempt to break new ground in understanding addiction redirects us to Heidegger's novel characterization of human existence, which he proposes at the beginning of *Being and Time*. We will discover that his new insight into what it means to be human brings to light, by way of contrast, a derivative viewpoint of the self that is emblematic of the false sense of identity of the addict in equating him/herself with the source of immediate gratification of the within-reach (e.g., the "substance"). Heidegger's claim that "the essence [what is most distinctive or "ownmost"] of Dasein lies in its existence" occurs in sharp contrast to the addict's misconception of his/her self-identity, such that, if could be articulated in philosophical terms, would be rendered as "The essence of the self lies in its *substance*."[2]

7.1 Priorities and Choices

Before there was gambling over the Internet, there were casinos. There were opium dens before ghettos became a place for the distribution of drugs. There were habitual drunks in the United States before we actually described them as "alcoholics." And yet since Bill Wilson's time, addiction in all its myriad forms has become a greater problem, not a lesser one. Does this occurrence reflect the limitations of the spiritual program of AA, or rather society's reluctance to come to grips with the growing problem of addiction? If Bill Wilson's Twelve Steps provide a blueprint for the practice of sobriety, just as the United States Constitution offers a blueprint for the pursuit of justice, then our understanding of the former must evolve with our changing times just as in the case of the latter. Rather than a fixed skeleton, could the Twelve Step Program, like a political constitution, include its own seeds of

[2] Heidegger 1977/1962, p. 56; tr. 67.

development to address the changing face of addiction within our fast-paced technological world?

When Wilson developed the Twelve Step Program, he devised a treatment that recognized the fine line dividing those who use alcohol in moderation from those who over indulge in a self-destructive way. In the intervening decades, however, most scientific and clinical studies of addiction tend to isolate discrete behavioral and personality traits that the addict exemplifies, e.g., deception and low self-esteem.[3] While this approach certainly has practical value for the professional physician and psychotherapist, perhaps additional light can be cast on the problem of addiction by proceeding from an opposite premise. Rather than isolate discrete characteristics, we might consider the *basic conditions of our human existence* that can reciprocally tighten or loosen, sharpen or blur, the boundaries separating the addict from the non-addict. Such an approach or methodology would allow us to see that the dependency of addiction can target virtually anyone and everyone; and, conversely, that just as we may always be presented with the prospect of falling into the trap of addiction, we are equally presented with the *possibility* of *not* doing so as well.

The fact that many individuals can drink in moderation shows that the substance (e.g., alcohol) does not simply cast a magical spell over the individual, even though the "fetish" character of the attraction seems to suggest as much. Yet, our earlier allusion to fetishes resumes a special importance here by helping to differentiate the "signifying" source of the arousal from the physical basis of dependency. In other words, that on which the addict may "depend" also evokes a chain of significations that connect other items of interest in the environment and, ultimately, point back to the self's possibilities. Thus, the source of addiction "stands for" something, which not only monopolizes the individual's interests, but also becomes a focus of complete self-absorption. In this case, whatever "addicts" not only endows the object/activity with inordinate significance, but also narrows the horizon within which all the other possibilities of existence emerge and become meaningful. Thus, the individual becomes forsaken or lost in pursuing his/her addiction, to the extent that it becomes the center around which everything else becomes meaningful, the new axis of a false self-understanding.

It is the complete reorientation of the individual's whole frame of reference to the substance/activity, and to it alone, which not only spawns but also *sustains* the addiction. That frame of reference includes the variety of ways and possibilities by which the individual stands toward him/herself, others, and the world at large. When a substance, e.g., alcohol, or another activity, e.g., gambling becomes the focus of one's entire frame of reference, *a preoccupation ahead of everything else*, such that all other concerns take a backseat or become secondary, then the circumstances for addiction are ripe. Those circumstances, however, are not held exclusively by the addict, but may potentially color any individual's situation. For that specific scenario arises when the preoccupation overrides the potential all human beings have to set priorities for their existence and to organize their lives. As a person becomes

[3] See Twerski 1997, pp. 68–69. See Nakken 1996, pp. 78–79.

addicted, the priorities that give direction to her/her life do not simply shift or become "skewed." Instead, these priorities are "inverted," in such a way that what is "first" becomes "last," and what is "last," the pursuit of recreational thrills, becomes "first." Within the context of our finitude, these priorities are the signpost and guidelines, which give direction and structure to an individual's existence, and by which he/she can exercise "self-control." When these priorities are inverted, however, then so too, the relentless pursuit of "what addicts" supplants and subverts ones' capacity to exert self-control, thereby enslaving the addict to the impulsive cravings of his/her desires.

As we have seen, however, addictions do not emerge from anywhere, without warning, but develop against the backdrop of the range of pursuits and activities that shapes the context of one's being-in-the-world. The undercurrent of our daily life suggests that the individual must in some way display a propensity or power, whether in the most simple or complex way, of allocating of what can be done in an interval of time and what can be accessed within an allotment of space. While the latter pertains primarily to the organization of our environment, and the former to the prioritizing of the self's possibilities, the interplay of time and space measure the boundaries and provide the stage on which the human drama, including the struggles of the addict, unfolds. When we consider addiction in terms of its unique dynamics, as well as the discrete characteristics that the so-called "addictive personality" must embody, we must consider the unique way in which time and space configure the platform on which the addict wrestles with the possibilities of recovery and loss. Though the field of psychology addresses the symptoms of addiction, how an individual manages his/her time and allocates his/her space, may tell us a great deal about the dynamics of addiction. Time appears the more obvious factor, because how one manages it casts considerable insight into a person's decisional processes. As is somewhat obvious, the interval in which one seeks to fulfill a desire, implying either delayed or "immediate" gratification, serves as an important measure of the degree of compulsiveness that drives one's addiction. Moreover, as Kemp emphasizes, the downward spiral of addiction resembles a chaotic vortex, disrupting the body's naturally based rhythms of sleeping, eating, and exercising.[4]

We cannot underestimate the importance of space. For any decisions pertaining to one's own welfare and that of others can never be separated from the confluence of environmental variables in which all of us carry out our daily lives. Foremost among these environmental factors may be the spatial proximity of the substance or "activity," in question, including, as in the case of illegal drugs, the distribution channels between the "pushers" and the "users." Through their dynamic interplay, time and space illustrate the change in a person's entire frame of reference in which the opportunity to become addicted becomes prominent. But what about the scenario of recovery; can the distribution of space also assume a positive role, as it does negatively by increasing the "availability" of the "within-reach" of what can become addictive? In order for the individual to re-set his/her priorities, he/she must begin to *inhabit a new space* of nurturing relationships: first, by associating with, e.g.,

[4] Kemp 2009, pp. 1–18.

even in the simple everyday manner of "hanging out," with constructive friendships and, secondly, by developing a therapeutic context of *proximity to others equally committed to the task of recovery*, e.g., through the "fellowship" of meetings. In the latter case, such meetings create a "safe-space" in which diverse individuals can exchange stories, expressing feelings, and, begin to "practice' the Twelve Steps. The safety in question creates a place (*Ort*) of disclosedness (*Erschlossenheit*) where individuals can be forthcoming, make amends, and find sanctuary (without fear of condemnation) while interacting with their peers.

But what characterizes addiction as a dramatic change in the texture of all one's activities and relationships? Two reasons stand out. First, it is important to rediscover the *context* in which addiction originates. That means addressing not only the various relationships that are affected by one's addiction, but also recalling the basic life-questions that arise when an individual struggle with his/her various demons even to the point of walking the tight-rope of life and death. These life-questions speak to the direction and purposes of one's existence, and are as equally as relevant to the addict as the non-addict–even though they frequently recede into the background of the hustle and bustle of everyday life. Whether when cast in a spiritual light or not, in one way or another the enigma of meaning touches us all, and that may even be more the case for the addict where a sense of purpose may all too easily dissolves into despair. And this is true, if only in reverse, in the case of the addict, where in dire circumstances he/she experiences a vacuum of meaning. In this regard, we cannot overlook the fact that "what addicts" has a significance in its own right, as it becomes a vessel, as it were, in which the individual disvests the entire meaning of his/her existence.

Second, we need to consider more closely the origin of the compulsive tendencies, not only as they translate into destructive behavior, but also as they distort one's self-understanding and ability to make decisions. When we speak of decisions, we are not simply separating them into moral categories of "good" and "bad," for either can apply to any of us. In most cases, we make choices among a different range of options and possibilities, even in the case of something as mundane as drinking Coca-Cola over Pepsi. Choice, then, exhibits a double dynamic in which the narrowing down of options precedes from a prior openness toward the range and diversity of them. But in the case of a compulsion, the tension of this dynamic weakens, and the entire focus of the decision-making process inverts. The openness of choice becomes secondary, and instead all the power of choosing is fixated on, or absorbed into, a single option. Any extreme compulsion ultimately results in some fixation or other, and thus we see that fixating on this, to the exclusion of everything else, involves collapsing the openness of choice. Addiction, however, takes this tendency to fixate one step further: not only is there a fixation on this or that, but the interest in what one is fixated upon usurps all other interests to the point as to overrule the power that allows to make distinctions about what we choose. Desire is thereby transformed into craving.

This transformation shifts the entire field of human concern/interest to whatever can fuel/satisfy the craving of the addiction. But the act of "fixating" does not necessarily rise to the level of addiction any more, than we have seen, that a fetish does.

While a fetish may dangle the "hook of addiction," a prior choice to pursue the craving may still occur, thereby allowing *alternative* concerns to arise as well. We must emphasize this distinction in order to appreciate the ambiguity, for example, which surrounds so-called "sexual addiction." An individual may like sex so much as to suggest a fixation, which may still not, however, be necessarily addictive. Specifically, a fixation rises to the level of addiction when it stifles all other alternatives, and the addict then flounders in a vacuum where the openness of choice recedes. We thereby encounter the unenviable situation, which a well-known slogan of AA recounts: "One drink is too much and a hundred drinks are not enough."

The difficulty we encounter in defining an addiction evokes a counter question as to obstacles that are inherently involved in formulating any kind of definition. A famous Supreme Court Justice once remarked that he cannot define pornography, but can recognize it when he sees it. He actually expresses a sentiment, going back to the time of Socrates, that definitions are not as easy to provide as may first appear. In the case of addiction, the difficulty of doing so may be even more pronounced, because identifying exactly when or why a person falls prey to addiction, or even whether a person is an "addict" at all, may not be nearly as straightforward as in the case of pornography. Indeed, the most *insidious* character of addiction is its self-deceptive, self-dissimulating, self-concealing character. In this regard, the layperson grapples with a proverbial quandary of whether it is the "how much" of drinking that actually distinguishes the alcoholic, or whether there is another measure. Does consuming alcohol on a regular basis, even to the point of inebriation, necessarily make one an alcoholic? This question reverberates in the familiar refrain of the habitual drinker who claims: "The fact I drink a lot does not necessarily make me an alcoholic."

As self-deceptive as the above claim of the alcoholic may be, it indicates the impossibility of providing a quantitative yardstick to define "addiction." Conversely, an occasion of over indulgence does not simply make one alcoholic. While no single definition will suffice, we can at least seek a "working" or "operational" definition to employ in discussing the phenomenon of addiction, in such a way as to *disclose* something crucial about our *existential predicament*. A person becomes an addict not simply due to the use or even overuse of a given substance, but rather, through such self-indulgence, that a single interest monopolizes all others in a way that proves to be destructive. In this regard, the craving itself, and whatever steps one can take to satisfy it, takes priority ahead of all other interests and concerns, minimizing their importance to the point that their fulfillment becomes inconsequential. In Heidegger's terms, the dynamic of projecting new possibilities *dissipates in favor of an inauthentic or unowned mode of "wishing" or "wanting" to bring what is "not yet" a source of gratification immediate proximity or "within reach" to be actualized in the present.*[5] Literally, this dissipating of the dynamic of temporality shapes the basic "fall back" or default stance of the addict, who sacrifices the broader prospects of the future to actualizing as "soon as possible" a sense of euphoria that is palpably "not yet." When the addiction becomes pervasive, then its pursuit

[5] Heidegger 1977/1962, pp. 458–461; tr.

can be achieved only to the detriment of, and in explicit contradiction with, all other tasks, concerns, and goals that constitute the promise of a richer future. Thus, the desire to watch baseball may "interfere" with the rabid fan's ability to get to work on time, but does not require a complete preoccupation with that activity to the exclusion of his/her job itself. But for the alcoholic, it is not only the case that drinking may come first (on occasion, like baseball), but that everything else comes last.

However one might cash out the obsessive-compulsive elements of addictions, the overarching question still remains about the meaning and purpose of one's existence. We might refer this factor as the "meaning-index." In simplest terms, meaning is whatever speaks to the individual's overall well-being and discovery of his/her uniqueness. Yet because human beings are finite, not all concerns and interests can be fulfilled. Rather, some must be placed before others, and arranged in degrees of importance. We can describe the form of that arrangement, which we experience as the need to set priorities, as a kind of index. In Heidegger's terms, the personal realization that life is finite, and that the development of a certain set of possibilities precludes developing others,[6] suggests that ultimately death yields this index. "... once [the possibility] of death has visited, life's priorities rearrange themselves.... Death-inspired existential wisdom can impart courage, resoluteness, and thankfulness."[7]

In the everyday course of our lives, we normally set priorities, for better for worse, without necessarily thinking about the process itself, or what we are doing when we do it–that is, until life throws us a curve ball and we must re-evaluate our priorities. A golfer or baseball player who is in the "zone" pays no attention to the mechanics of the swing until something goes terribly wrong. Similarly, we do not address the overall profile of our well-being until we are faced with some kind of crisis; but enter the alcoholic, the addict. The addict epitomizes the extreme of what can go wrong in an individual's life, providing almost a mirror image of the meaning-index or exemplifying it in *reverse*. The character of the addiction is that the craving, and its gratification through a substance, becomes the false "stand in," imposter," or artificial substitute for one's overall, complete well-being. The reliance on a 'substance' represents the inauthentic, unowned temporalizing that rebuffs the futural challenge of "having one's being to be" in favor of the false security and complacency of a 'present' *stasis* of gratification.

Acquiring and using the substance, then, becomes the basic *modus operandi* of the addict, allowing that pursuit to speak in the name of one's overall well-being. The addict's craving absorbs all interests back into itself, thereby renouncing the discipline of the priorities and making the meaning-index all but obsolete. When the addiction becomes pervasive, the entire compass of human concern, including needs, commitments, and responsibilities, get sucked into the vortex of the craving and everything "is pressed in service of the addiction."[8] In the overturning and

[6] Heidegger 1977/1962, p. 378; tr. 331.

[7] Aho and Kevin 2008, p. 126.

[8] Heidegger 1977/1962, pp. 259–260; tr. 240.

supplanting of an individual's concern for his/her general well-being as a person, addiction reigns.

Here we arrive at an operational definition of addiction, which can be amended as we continue to fathom the complexity of the phenomenon. In this respect, it is neither (1) the behavior alone which constitutes the addiction nor (2) the dependence upon a substance, as manifested in the *symptoms* which arise from abruptly refraining from its use, e.g., "withdrawal." Rather, it is the wholesale *destructive* character of the habit, particularly upon oneself but indirectly upon others as well, which distinguishes addiction. In this regard, addiction is like flipping over the moral and spiritual "compass," which has previously guided one's life, such that the addict makes choices in precisely the opposite way he/she typically would have *chosen*. When the compass of one's existence points completely in the opposite direction, and one's priorities are completely inverted, the individual then teeters on the brink of self-destruction. Heidegger's allusion to the "downward spiral" (*Absturz*) of addiction[9] suggests precisely this scenario: the inverting and usurping of one's priorities on the one hand, and the relinquishing of self-control on the other, delivers the individual over, i.e., as in the etymology of the word "*addictus*," to his/her own "self-destructive" tendencies.

To be sure, it would be folly, and perhaps would bring a smile to Socrates's face (were he still alive) to argue for the superiority of this definition over any other. Nevertheless, we would be remiss not to emphasize the two key points that follow from this phenomenological experiment of defining addiction. First, our approach seeks to remain true to the phenomenon by returning to the root of the word itself, and thereby capturing the elemental simplicity of what today involves highly complex forms of behavior. Secondly, in honoring the "facticity" of experience in this way, we allow the simplest "formal indicators" to yield the vocabulary for addressing the phenomenon of addiction. In this way, a *new discourse for discussing the problem of addiction can emerge, which is completely different from those that shape the accounts in the alternative fields of medicine, psychology, and sociology*.[10]

Having circumscribed the parameters of this discourse, it is still necessary to establish what are the precise benefits for discussing the problem of addiction in phenomenological terms, rather than in other, more customary ways. Let us then outline what some of these advantages may be.

7.2 Responsibility and Recovery

In placing our discussion on an existential plane, we are able to address concrete issues that parallel the problem of addiction, but at the same time remain notoriously elusive. Foremost among these is the extent that one has freedom over both

[9] Heidegger 1977/1962, pp. 259–260; tr. 240.

[10] See Aho, James and Kevin Aho 2008, p. 150.

averting and surmounting an addiction, and, conversely, the extent of one's "responsibility" toward one's impulses and actions. The question of responsibility, however, brings many different issues into play, including the relevance that "morality" may have within this overall discussion. On the one hand, since no one as a human being can be completely immune from the possibility of addiction, there is not one of us who can stand on exalted moral ground. Indeed, the first of the Twelve Steps—the fact that individual is "powerless" over alcohol—underscores the importance of acknowledging one's frailty. On the other hand, there is an obvious difference between amorality, or moral indifference, and relinquishing the pretense of self-righteousness. Indeed, Bill Wilson clearly recognized a moral side to the process of recovery, insofar as purification and self-humility provide the cornerstone for the individual's restoring a sense of equilibrium in his/her life. Modesty in the face of one's affliction constitutes the first step on the road of taking responsibility. Peg O'Connor summarizes the moral quandary posed by the societal concern with the problem of addiction, while emphasizing the importance that a philosophical orientation has in contrast to prevalent psychological models:

> Pairing rich moral language with discussions of addiction is a delicate matter. For centuries now, addiction has been considered a moral failing; addicts where those who simply lacked the willpower and moral fortitude to stop. We suffered from weakness of will. Either we were totally impetuous or we overrode what our reason told us to do. Addicts were thought of as weak-willed, lazy, selfish, and guilty of a host of other blameworthy traits.
>
> With the medicalization of addiction a shifting tide toward explaining it as a "brain disease," the moral disapprobation has lifted some, and discussions about responsibility have shifted. This is clearly a very good thing. But blame and responsibility are only two thin slices of moral philosophy. Character and virtue belong in the moral realm and can illuminate many previously neglected dimensions of addiction and recovery.[11]

In this we regard, we must briefly consider a further addendum to the two points outlined above in regard to the meaning and importance of responsibility. For Heidegger, responsibility can no longer be equated with a volitional act or exercise of the will[12]; perhaps ironically, this view squares with the caveat of the Twelve Steps that the claim of the individual will must be sustained by, and subordinated to, a "higher power," in order that the recovering addict can develop both the humility and resoluteness to become responsible. Correlatively, for Heidegger, responsibility is a way of becoming "answerable," such that responsibility does not depend upon subjective predilection, but instead originates from the place, the niche, he/she occupies with the wider expanse of the openness and harmony with what is (e.g., "being" [*Sein*]). No responsibility can be achieved without the prior concession of humility, that is, of developing a new *life-trajectory* within the wider expanse of openness, of being itself.

---

[11] O'Connor, pp. 109–110.

[12] Willing can un-will its own subjective bias, its self-enclosedness, in order to enact a choice that originates from the openness. See Schalow 1993, 96–102.

To the extent we can express a moral perspective, perhaps it does not entail echoing society's disapproval of the addict, but, on the contrary, getting society to question itself, that is, allowing each of us as individuals to address the points where our institutions are in complicity by supporting addictive behavior. Today, we live in a society that has become increasingly depersonalized, particularly with the computer age and the accelerated advance of technology. It is not so much that the machines are the problem, but an accompanying world-view or mentality that makes usefulness, in regards to people as well as things, the end all and be all in the measure of worth. The difficulty with this mentality, however, is that it becomes easier and easier for *many of us to live in a grey area of blurring the lines between using people and using things*. For example, the multinational corporations in which many of us may work treat employees as substitutable parts, the proverbial cogs in the machine. The addict falls into the cracks and crevices where these lines converge: that is, living itself reduces to using, and every pursuit can be summarized by the simple formula such that the there is only one end, the satisfaction of the craving, and everything else serves as a means. Technology does not create the addiction, but instead perpetuates a climate of meaningless pursuits in which, as it were, the addict can find a niche of his/her cycle of addiction. *The "hook" of addiction is thereby reinforced and strengthened by the commodification of the fetish* and its self-deceptive practices,[13] which conceals or masks the destructive (e.g., narcissistic) roots of human desire. Ironically, the products of technology, by including the trappings of wealth, comfort, and social status, make it easier, at least for a time, for the addict to occupy this niche, as the example of many celebrities demonstrates.

What goes unnoticed, however, is that the addict implicitly endorses an amoral viewpoint in which the satisfaction of his/her craving becomes the end of all ends and everything else but a mere means. In this respect, we distinguish the attitude as amoral, in contrast to the evaluation of all accompanying behavior, e.g., lying and cheating to support the addiction, which in the eyes of ethicists would be deemed "amoral." Perhaps the greatest ethicist of the modern age, Immanuel Kant, advanced this basic moral principle: "Act always, so as to treat humanity, whether in your own person or in that of another, as an end only and never merely as a means."[14] Though we tend to emphasize Kant's prohibition against using others or exploiting them for the benefit of our own interest, the individual can be self-exploitive as well. Indeed, in cases where addiction becomes prevalent, the addict submits to the dictates of the craving in such as way as to sacrifice his/her health, job, and most of all, his/her dignity—not to mention his/her friends and family—for the gratification which the substance provides. The diminishment of moral self-awareness, even to the detriment of the addict him/herself, tells us some unique facts about the problem of addiction: the seemingly paradox that when the "me" of self-interest reigns supreme, the individual has started down a destructive path.

---

[13] For further discussion of this dissimulating pattern or gestalt, see Lefebvre 1968, pp. 47.

[14] Kant 1981, p. 36.

Once again, Kant's distinction between "morality" (*moralitas*) and "legality" (*legalitas*) " may prove helpful here.[15] Indeed, the phenomenon of addiction provides an interesting example of Kant's claim that morality is always initiated from within the individual, a "motive" inwardly cultivated, in lack of a better word, rather than externally imposed or held out an incentive, punishment, and reward. For Kant, this inwardly motivated commitment to morality takes precedence over the assessment and evaluation of the *consequences* of one's actions. This does not mean that the latter is not important, but primarily from a "legal," rather than a "moral" standpoint. Obviously, society must hold the addict accountable for the consequences of his/her action, e.g., driving while intoxicated, including incarceration if necessary, and, in failing to do so, further "enables" the individual. But as Steiner points out, admonishments, threats, and even the fear of retribution frequently have little effect on changing the addict's behavior.[16] This simple fact testifies to the truth that change, like morality, is ultimately motivated from within, rather than from without. Moreover, the bonds of co-dependency must also be broken, in order to provide a ripe climate for such a dramatic change or self-transformation.

Kant recognized that freedom was a prerequisite for morality. The source of this freedom, however, remained for him somewhat of a mystery. Heidegger, who more than a century later appropriated much of Kant's philosophy, addressed the question as to the origin of freedom.[17] Specifically, Heidegger saw that freedom is a power that the self displays when it uncovers its own possibilities, and hence includes "moments" of discovery and disclosure, as much as choices involving the exercise of the will. Thus, freedom transposes the self into a wider expanse of openness, when it can become receptive to alternative possibilities, thereby relieving the tendency to become "fixated" on any one thing or pursuit. From this openness arise new levels of self-understanding, allowing the individual to find his/her own path or direction in life, and thereby, conversely, avoid the temptations, compulsions, and pressures (including of one's peers). Through such openness, we can become as much "liberated" by our desires as overtaken by our narrower cravings. In the former case, the self (through its embodiment) redirects or maps its desires back upon the wider expanse of its possibilities, including the impetus to seek its *own* "highest aspirations" or "transcendence." As Lawrence Hatab states:

> Authentic desire, then, could follow from being educated in the finitude of being, where one can overcome the "addiction" to presence and actuality that prompts a flitting about in search of perpetual fixes of surface stimulation, acquisitions, and consumption. Such addiction is not only an alienation from the meaning of being in general, but also from a deeper sense and appreciation of specific phenomena in the world. [18]

---

[15] Kant 1996, p. 17.

[16] Steiner 1971, pp. 50–53.

[17] See Schalow 2013, pp. 129–133. In the early 1930s, Heidegger was on the way to linking freedom with openness, and ultimately to "letting be" in light of Meister Eckhart's concept of "releasement" (*Gelassenheit*).

[18] Hatab 2000, p. 127. As Hatab emphasizes, Heidegger distinguishes between urge (*Drang*) and authentic care (*echte Sorge*), which along with inclination (*Hang*) are facets within the total structure of care—vis-à-vis its manner practical comportment in the world.

As a precursor to a sense of responsibility or even a (moral) "virtue" of self-directedness and restraint, authentic desire elicits the individual's potential *to be a self.*

With this owned dimension of free engagement arises responsibility, if by this we mean not simply "accountability," but rather "answering" to a prior claim to be true to oneself and thereby take the initiative to be an individual. Because concealment and denial fuel addiction, and are contrary to an individual's becoming "answerable," the prospect of assuming responsibility remains most completely foreign to the individual. Indeed, the undeniable paradox is that the impetus toward responsibility is what the addict possesses least, but needs most—a "double lack and a double not," to paraphrase from Heidegger's discussion of Hölderlin's poetry.[19] Herein lies the ontological "fact" as to why it is so difficult for the addict to break the cycle of addiction and start on the path of recovery.

7.3 Choosing to Choose: Transcendence as a Pathway

The transition to recovery is a process, a journey, or, in Heidegger's terms, a pathway. All journeys begin with a single step, and this is never so true as in the case of recovering from addiction. But prefiguring the obvious step (albeit negative decision) *not* to use, to refrain from self-indulgence (and pursue sobriety), is the act of choosing out from which one's authentic temporality flows, that is, *choosing to choose*. Herein lies the origin of authentic temporalizing, which encompasses the entirety of one's existence and yields the momentum whereby the individual *can* reverse the inertia of his/her fixated desires. By this inaugural act of choosing to choose, the individual takes the first step along the path of self-transcendence, i.e., developing a new life-trajectory. As Heidegger emphasizes, the first level of choosing is the choice to be oneself, or "resoluteness" (*Entschlossenheit*). In this resoluteness, the self chooses from an appreciation of, and humility before, the abundance of possibilities, rather than "fixating" on the immediate opportunities for gratifying one's desires.

A more fundamental "desire" as the "motivation" of open responsiveness prevails over the narrow "craving" of satisfying one's urges. In an analogous, but more radical way than Kant, Heidegger realized that freedom prompts "self-directed" initiatives, rather than constraints imposed by others or even the compulsion of one's desires. Just as morality cannot be imposed from "without," so the impetus, motivation, and initiative to change, from the standpoint of the individual who has become "enslaved" to his/her cravings, must also be "self-induced." But, in the case of the addict, does this "change" involve simply a choice to indulge or not to indulge, e.g., to drink or not to drink? If only the choice were this easily made. The difficulty here does not lie so much with the sentiment, but rather with the understanding of choice. The power to choose it not simply a function of the will, like a switch that

[19] Heidegger 2000a, p. 64.

can be turned on and off. By the same token, the individual can be empowered to choose not to indulge; the question is simply how that empowerment can occur. The basic thrust of the treatment of AA, of course, is to appeal to a higher reality as the source of that empowerment, however we define it. Nevertheless, the "empowering" that returns to the individual his/her capacity to choose, occurs only through a process of complete transformation. Because within the grips of addiction the individual's cravings monopolize all his/her concerns and pursuits, his/her recovery can only occur, then, as a complete reversal.

All that one has "divested" of oneself in pursuing the addiction must in turn be reinvested, as it were, back into the development of who one is, that is, the search for his/her individuality. This dramatic re-orientation of one's existence, of course, involves renouncing "what addicts." The simple act of renouncement, however, is not a definitive "cure," a climax or a culmination, but instead constitutes a new beginning, a point of departure along a journey. As in every case, the journey begins with a single step, and hence with the humility of acknowledging its lengthy progression. In Heidegger's terms, the humility is an admission of one's finitude and equally the circumstances of one's facticity. The facticity is such that while the initial "decision" to "stop using" is abrupt, the progression along the new path is commensurately slow. As O'Connor states: "The commitment to recovery involves fundamentally altering how an addict *is* in the world. This cannot be done wholesale but only incrementally."[20] Hence, the actual enactment of the decision is inversely proportional to its initial abruptness. One of the chief maxims of AA, then is "one day at a time." In other words, the facticity of human circumstances being what they are, the only interval of time that can reasonably exhibit the individual's fortitude to enact his/her decision is the length of a day.[21] This basic "mantra" of recovery proves to be efficacious as a guideline of treatment,[22] precisely because it approaches a life-crisis as it originates from its deepest root in human temporality, the self's unownedness of seeking immediate gratification in the present by escaping into an illusory future and a fugitive past. The strategy of recovery, in a manner commensurate with the plight of addiction, presupposes a dynamical view of the self. According to these dynamics, the self experiences its freedom by acknowledging those limitations that are set through a finite allocation of its *own* temporality: that one's power to refrain from addictive behavior, e.g., drinking, does not extend beyond the "day" and hence must be appropriated again each day forward.

The Twelve Steps and the various maxims for implementing them, which have evolved over decades, are neither occult incantations nor empty slogans. Yet, their effectiveness lies in the individual's ability to incorporate them as guidelines, which translate into the dramatic change both in attitude and action. Their truth lies in how they disclose to the individual a new path of existence, or way of living, apart from the reliance upon and use of "what addicts." Yet, the "outsider" tends to remain

[20] O'Connor 2016, p. 132.

[21] Viktor Frankl makes this point in emphasizing the therapeutic value of the individual assuming responsibility, even from moment to moment. Frankl 1980, pp. 64–65.

[22] M. Valverde 1998, pp.136–137.

skeptical, if only by mistakenly evaluating the Twelve Steps, not in terms of their efficacy but rather as presenting certain theoretical claims or assertions. In developing the Twelve Steps, however, Bill Wilson's message was "to keep it simple."[23] With this keynote, he emphasizes that his program served as a practical manual, a road map for a journey, which, while its first step leads away from addiction, *opens one up to the greater adventure of life*. Practice over creed, praxis over theoretical models, would then describe the existential thrust of the Twelve Step Program.

In one sense, we might perceive an intentional vagueness in the formulation and application of the Twelve Step Program. Indeed, its very design, as its creator emphasized, was to allow enough leeway for the individual to tailor it to meet both the uniqueness of his/her own circumstances and the level of his/her self-understanding. Thus, what initially appears as "vagueness" involves shifting the focus away from the rigidity of any creed or dogma to extending an open invitation to the individual to author his/her personal "life-story." In other words, the slogans of the Twelve Step Program do not function as specific claims standing on their own, but instead serve as "signposts" along a lengthy path of transformation, which the individual must travel in order to break the shackles of his/her addiction. Because the patterns, indeed "life-styles" of the addict may be cultivated over a significant temporal span, we can expect that the corresponding transformation has its own "course" and development, as it were, within the overall "rhythm" of life. The incremental steps that are required to alter a chronic form of habituation do not occur in a vacuum, however, but must also include modifying other, if albeit, secondary life-habits, including nutrition and exercise. Changes in the self's factical circumstances, including the most basic bodily reflexes and responses, help to reshape the overall tapestry of care and thereby contribute to developing a new life-trajectory. As David Morris emphasizes in recasting the problem of addiction in terms of a dialogue with Hegel's philosophy: "This is the experience that changing our own action is the work of a lifetime, that change of action is not local to the here and now. Changing a habit–and thence ultimately acting in a way that is of our own choosing–requires a sense of time other than the one supplied by a purely rational world-view."[24]

In light of Heidegger's hermeneutics, to appropriate is to make one's own. And this way of making one's own originates from, and makes explicit, the unique and singular path of one's journey throughout, which illuminates "who" one is, i.e., from the depths of his/her "ownmost," authentic potential.[25] We thereby speak in terms of "coming into one's own," as when a person, young or even old, finds his/her path in life that holds the promise of yielding the greatest possibilities, i.e., for

---

[23] One of Bill Wilson's reservations with the Oxford Group (see Chapter Five) was that such lofty ideals as the "Four Absolutes" were beyond the reach of the so-called average alcoholic, and hence broke with the simplicity of the Twelve Step Program. Kurtz 1991, p. 51.

[24] See Morris, 2001, p. 393.

[25] Without addressing the philosophical roots thereof, there are recent attempts within the popular literature to identify the discovering of the "authentic self," e.g., or "finding out who one really is," as the cornerstone for the recovery from addiction. The terminology of authenticity has contribution to shape the attempt to conceptualize the problem of addiction. See Lawford 2014, pp. 7–8.

"fulfillment" of his/her potential. Ultimately, then, *the catalyst for the transforming power of the Twelve Steps, as a practical manual or blueprint of life, lies in the "free-initiative" of its appropriation and "ownership" by each and every individual.*

From a phenomenological perspective, the genius of the Twelve Step Program lies in basing its efficacy on a personal challenge posed to an individual, as much as extracting the kernel of Christianity. Indeed, the success of the program lies in its exercise as a form of "praxis, " which calls the individual to take the initiative to change, and thereby summons him to act in a manner to fulfill this commitment. Because addiction normally involves a process of habituation, only through a proportionately lengthy period in which the individual follows through on his/her commitment to change can a lasting transformation occur. In this respect, much of the mystery surrounding the Twelve Step Program points back to the depths of the individual's ownmost openness to, and exercise of freedom.

Recalling the root of the word, *addictus*, as being delivered over to, dispossessed, and divested of one's individuality, we see, that, by contrast, where the zero-point of all "treatment" begins: namely, in the initiative (i.e., the freedom) the individual shows in answering the ownmost appeal to one's individuality. Whatever therapeutic practices one may follow, no formula for breaking the shackles of addiction can occur without the individual's initiative. Even when others stage an "intervention," only an occasion is provided to help the addict. But the "leaping-off point" (*der Sprung*) of recovery is still the initiative of the power to choose. Although not the only philosophical approach illuminating this dimension of the self, Heidegger's thinking certainly offers an insightful glimpse into the possibility of healing and recovery.

As we bring this chapter to a close, let us consider how our attempt to break new ground in our understanding of addiction redirects us to Heidegger's novel characterizations of human existence, which he proposes at the beginning of *Being and Time*. Specifically, his new insight into what it means to be human brings to light, by way of contrast, a derivative viewpoint of the self that is emblematic of the false sense of identity of the addict in equating him/herself with the source of immediate gratification of the within-reach (e.g., the "substance"). Heidegger's claim that "the essence of Da-sein lies in its existence" occurs in sharp opposition to the addict's misconception of his/her self-identity, such that, if could be articulated in philosophical terms, would be rendered as "The essence of Dasein [as a subject] lies in its *substance*."[26] Within the context of our study, the term "substance" (that is put in italics in the preceding quote) assumes a double meaning: first, in a reference to a derivative concept of the self (as subject) as something "present-at-hand," as an enduring stasis falsely emulating eternity (and hence as reflecting the metaphysical tradition since Descartes by dividing the human being into mind and body, thinking and extended "substance"); and, secondly, the everyday characterization of a primary hook of addiction as a "substance," (e.g., alcohol or drugs). In the stasis of the addict's comfort-zone, the fetish-like craving of addiction comes to light as a

[26] Heidegger 1977/1962, pp. 421–422; tr. 365–366.

narcissistic attachment. The narcissistic desire for immediate gratification, the manufacturing of a false sense of eternity ahead of the limitations of our temporal-spatial "earth-boundedness," occurs at the expense of authentic solicitude or reciprocity of *love*.[27] When oriented in terms of finitude, the *eros* of human desire can fuel the self's striving to be authentic (rather than ontically fixated on an narcissistic infatuation). Finite transcendence defines the arc of self-surpassing, which propels the self into the wider expanse of openness. The act of transcending, or the impetus of "ontological desire" (vs. ontical craving), elevates the individual toward those higher possibilities that speak to his/her individuality, i.e., over and beyond the narrower reach of its immediate impulses.[28] As such, transcendence is the original source of ecstasy which leads to the self's affirmation of life and its limitations, rather than to the denial of the latter in a narcissistic search to fuel the insatiable desire for gratification. By making this contrast explicit at the most basic level of our inquiry, we arrive at the key interpretive guideline for developing a *language* to address the *phenomenon* of addiction. Specifically, an existentially nuanced language speaks to the larger life-drama of the human predicament, rather than characterizing addiction as a strictly behavioral (or even neurological) condition.

7.4 Conclusion

The advantages of developing a different, more dynamically and existentially "charged" language by which to discuss the problem of addiction, have become apparent. Accordingly, we can now give deeper expression to the basic maxims of the Twelve Step Program, opening up a wider horizon in which their meanings can be understood and appreciated anew. Given this new circle of understanding, let us then revisit the possibility of "recovery" and outline its key elements.

References

Aho, James, and Kevin Aho. 2008. *Body matters: A phenomenology of sickness, disease, and illness*. Lanham: Lexington Books.

Boss, Medard. *The meaning and content of sexual perversions*: *A daseinanalytic approach to the psychopathology of the phenomenon of love*, trans. Liese Lewis Abell. New York: Grune & Stratton. 1949

Frankl, Viktor. 1980. *The doctor & the soul. From psychotherapy to logotherapy*. New York: Random House.

---

[27] Boss 1949, p. 37.

[28] Zimmerman 1995, pp. 511–512. "Reading Plato's *Eros* as *Seinserstrebnis*, ontological desire or striving-after-being, he [Heidegger] asserted that to be human means to desire being [*Sein*], to be erotically drawn toward it, to be moved into existing, transcending, opening up the temporal-historical world" (p. 511). Also see Schalow 2001a, pp. 88–89.

Hatab, Lawrence J. 2000. *Ethics and finitude. Heideggerian contributions to moral philosophy*. Lanham: Rowman and Littlefield.
———. 2016. The hurdle of words: Language, being, and philosophy in Heidegger. In *Hermeneutical Heidegger*, ed. Michael Rowler and Ingo Farin, 262–282. Evanston: Northwestern University Press.
Heidegger, Martin. 1962. *Being and time*, trans. John Macquarrie and Edward Robinson. New York: Harper & Row.
———. 1977. *Sein und Zeit*, GA 2. Frankfurt am Main: Vittorio Klostermann.
———. 2000a. *Elucidations of Hölderlin's poetry*, trans. Keith Hoeller. Amherst: Humanities Press.
Kant, Immanuel. 1981. *Grounding for the metaphysics of morals*, trans. James W. Ellington. Indianapolis: Hackett Publishing.
———. 1996. *The metaphysics of morals*, trans. Mary Gregor. Cambridge: Cambridge University Press.
Kemp, Ryan. 2009. The temporal dimension of addiction. *Journal of Phenomenological Psychology* 40 (1): 1–18.
Kurz. 1991. *Not-God. A history of alcoholic anonymous*. Center City: Hazeldon.
Lawford, Christopher Kennedy. 2014. *What addicts know: 10 lessons from recovery to benefit everyone. Foreword by Drew Pinsky*. Dallas: BenBella Books, Inc..
Lefebvre, Henri. 1968. *The sociology of Marx*. New York: Random House.
Morris, David. 2001. Lived time and absolute knowing: Habit and addiction from 'infinite jest to the *Phenomenology of Spirit*'. *Clio* 30 (4): 409–412.
Nakken, Craig. 1996. *The addictive personality*. Center City: Hazelden.
O'Connor, Peg. 2016. *Life on the rocks. Finding meaning in addiction and recovery*. Las Vegas: Central Recovery Press.
Schalow, Frank. 1993. Willing as the genuine postscript of modern thought: At the crossroads of an anomaly. *Epoché* 3: 77–104.
———. 2001a. Eros and the God-question. The crossroads between hermeneutics and psychoanalysis. *Review of Existential Psychology & Psychiatry* 25: 79–91.
———. 2013. *Departures: At the crossroads between Heidegger and Kant*. Berlin: de Gruyter.
Steiner, Claude. 1971. *Games alcoholics play*. New York: Grove Press, Inc.
Twerski, Abraham. 1997. *Addictive thinking*. Center City: Hazelden.
Valverde, M. 1998. *Diseases of the will: Alcohol and the dilemmas of freedom*. Cambridge: Cambridge University Press.
Zimmerman, Michael E. 1995. Ontical craving versus ontological desire. In *From phenomenology to thought, errancy, and desire: Essays in honor of William J. Richardson*, ed. Babette Babich. Dordrecht: Kluwer Academic Publishers.

Chapter 8
From Excess to Economy: The Elements of Recovery

The medical practice today reflects the world-view of modern technology—of both physician and patient alike—in the expectation to package various "cures,"[1] including through the use of prescriptive drugs. Included in this world-view is a presumption of progress, and, indeed, a linear vision that as technology advances, so does the quality of human life. But as medicine has vanquished many of the "external" threats of a century ago, and thus significantly extended life-spans today,[2] our vulnerability to so-called "internal," emotional frailties have increased in inverse proportion. The term "neurasthenia" distinguishes the "emotional flatness and exhaustion from a life increasingly mediated by the mechanized acceleration and time pressure of the industrial age."[3] There may be sociological and anthropological explanations accounting for this development: including urbanization, the decline of the family, impersonalized life-styles, and other forms of uprootedness such as threats of global terrorism. Yet, philosophically speaking, the development of a narcissistic culture of self-absorption, on the one hand, and, on the other, the presumption of progressively higher standards of living, e.g., resulting in happiness, betray an illusory premise. That is, liberation from material want insulates the individual from the challenges and pressures of the human predicament. As such, the proliferation of stress is not simply a medical symptom to be clinically diagnosed, but instead is a sober reminder of our fragile place in the universe and a thirst to find meaning in our lives.

When viewed from this broader, philosophical perspective, the plight of addiction likewise ceases to be only a medical condition. On the contrary, our study throughout reveals that the human vulnerability to addiction is instead a signpost to

[1] In *Zollikon Seminars* remarks (Dec. 18, 1963) that it would not do harm for "physicians to have something of Kant in their libraries," implying a scientific-technological bias on the part of the medical profession." Heidegger 2001, p. 267.

[2] See Kurzweil 1999, pp. 234–239.

[3] Aho 2009, p. 105. As Aho points, out the physician, George M. Beard, coined this term in 1881.

F. Schalow, *Toward a Phenomenology of Addiction: Embodiment, Technology, Transcendence*, Contributions To Phenomenology 93,
DOI 10.1007/978-3-319-66942-7_8

a what most aptly describes a sense of "self-disconnectedness" experienced throughout much of humanity today despite, the rewards and comports provided by modern technology. What is stress, then, but the pervasiveness of this experience as reflected in the popular lexicon, e.g., "I am stressed out...." But what do such common locutions suggest other than the possibility that, when the pressures become too overwhelming, and the ability to "cope" becomes proportionately too ineffective, the possibility of "falling prey" to addictive behaviors may be imminent.

If we consider stress as an acronym for our vulnerability to addiction, then its vagueness and indefiniteness suggests more than a medical symptom. Instead, while there still may be medical prescriptions and remedies, herein lies an example where if ever were such a need for one a "holistic" approach or form of treatment is in order. If we consider the plight of addiction is this "holistic" way, however, wellness cannot be found in a "bottle," since the reliance on such (as we refer in the vernacular to a container of alcohol) is part of the problem. Indeed, whatever medical remedies figure in to the overall plan of treatment, these cannot be separated from the *development of an individual's unique life-trajectory* and its relevance for addressing the cultural climate of stress. The same can also be said for various clinical approaches, including talk therapy and even group therapy as may occur under the auspices of a network of peers or "fellowship." The fact that treatment may occur through various channels, and traverse different levels of personal development—spiritual, physical, and psychological—only reinforces the importance of re-situating the problem of addiction within this wider philosophical horizon. If our aim throughout has been to develop a philosophical understanding of addiction, we must now consider whether our study has reaped any fruits on a practical level as well, that is, by shining light on new paths of recovery.

In considering the possibility of treatment there is an obvious presumption that addiction is, after all, a problem that is treatable. But once again, our philosophical inquiry leads us to pause before the obvious, and recognize instead that, in terms of the temporalizing of the self, treatment must occur *on a pace* that is completely contrary to the quick-fix mentality of our culture, on the one hand, and, on the other, the addict's entrapment in the specious present of immediate gratification. The common-sense acceptance of this belief is certainly not unreasonable, particularly given the relative success of the Twelve Step Program and the proliferation of rehabilitation centers around the world. But what if addiction were a problem that metamorphosed under its own impetus, outstripping the conventional models of treatment? What if the problem were analogous to antibiotic-resistant staph infections, which, while treatable, highlights how advancements in our technology can spawn crises that are not always solvable by the same ingenuity, practices, and devices? Although this analogy should not be taken too literally, it may suggest a possible *paradigm shift* in how we view addiction. Thus, the danger of technology lies in fostering the illusion that left to its own devices a remedy may magically occur, thereby masking the paradox that the temporal enactment of recovery occurs at the expense of immediate gratification. Perhaps the realization that technology drives human compulsiveness—the *willing of our self-willfulness*—points to a turning point in civilization in which the exploitation of nature, on the one hand, and the (self-)victimization of humanity, on the other, are not two separate occurrences, but instead converge at a single crossroads. If this were the case, then the problem of addiction would be emblematic of a crisis rippling across all corners of the globe, and thus provoking the most perennial questions of who we are, the meaning of life, and even our role as custodians of the earth. If only by helping us to see the *larger picture*, the philosophical approach to addiction occupies a special niche in the ongoing discussion of this vexing and ubiquitous problem.

The so-called "paradigm shift," however, is not automatic, if only because of the misunderstanding—both cultural and individual—which continue to pervade the problem of addiction. These misunderstandings are in turn reflected in how the various "treatment" programs continue to oscillate between "spiritual" practices (e.g., from the Twelve Step Program to so-called "talk" therapy) and medical-clinical procedures (e.g., the inducement of changes in brain-chemistry through prescriptive medications). From a philosophical perspective, this oscillation is symptomatic of a deeper schism in thinking and conceptualizing of which we are the heirs today, namely, the mind-body dualism whose influence lurks in the background of most "theoretical" models for treating addiction. The hermeneutic-phenomenological method implemented throughout this study circumvents this dualism; and, in the process, suggests new avenues for developing treatments which are not restricted to the narrow presuppositions of either side of the dualistic bias, e.g., either skewed in the direction of mind or body. While proceeding from a different philosophical orientation, O'Connor arrives at a similar conclusion: "Most recently, neuro- and brain scientists are entering the fray, hypothesizing that addiction is a chronic biochemical condition. But here, too, the questions about the relationship between the mind and the body pop up anew and still resist obvious answers."[4] Indeed, psychopharmacology as a form of therapy, while useful in certain contexts, rests on two basic assumptions: first, a mechanistic world-view in which the patient's illness can be treated symptomatically without addressing deeper life-questions and, secondly, a technology model of the health-care industry, which imposes standards of efficiency and the most cost-effective, expedient treatments only to reinforce the fast-paced, frenetic pace of living in today's society.[5]

Yet, however compelling this philosophical perspective may be, its importance resides in offering new "clues," if not in outlining the key elements of any "program" of recovery. Rather than view that recovery in a linear way of moving from step A (of using) to step B (not using), we might construe the process as a circular one, which revolves within a new orbit of self-awareness. This process would in turn correspond to elliptical arc of temporalizing that provides the crux of the development of twentieth century phenomenology (including the application of Heideggerian phenomenology in the therapeutic practices of Medard Boss). The "end-game" would no longer be a search for a "cure," which addiction specialists going back to Bill Wilson all disclaim. Instead, breaking the compulsive behavior would both proceed from and return to the dynamic of broadening the reach of one's

[4] O'Connor 2016, pp. 5–6.

[5] Aho 2009, pp. 120–121.

self-awareness to include a concern for others, one's place within society, and on a deeper level an appreciation of an affinity or kinship with all things (even the realm of nature and animals).[6] Our aim throughout has been to recast various programs of treatment within the wider context of the temporal-spatial enactment of human existence and the rediscovery of the self within the larger compass of the human predicament. As mentioned in the previous chapter, the therapeutic remedy of dialogue through a setting of "fellowship," even broadly speaking "group therapy," presupposes the allocation of spatiality at a fundamental level of human interaction. Thus, the creation of a safe-space (or a "safe-*place*") facilitates the healing process by allowing the strength of a bond with others (e.g., *Mitda-sein*) to support each individual to confront the darkest recesses of his/her struggle for recovery. The allocation of lived-space in both avoiding the negative social and environmental influences, e.g. the debauchery and "peer pressure" of saloons, and in cultivating constructive relationships, e.g., through a "fellowship" of sharing, goes hand in hand in the self's development of a new temporal arc or life-trajectory. Indeed, the rediscovery of the self can occur within the temporal-spatial openness (*Zeit-Spiel-Raum*) of its inhabitation of a world, including its being-with others.

When the individual ceases to view his/her life in a narcissistic struggle (in opposition to everything else), and rather as belonging to and inhabiting this wider expanse of being, then self-awareness (to employ Heidegger's term) changes into *mindfulness* (*Besinnung*). In the deeper sense in which Heidegger envisions, the mindfulness is both a "release" from the compulsiveness of technological control and simultaneously a "releasement" to a measure or sense of proportionality pervading all things. As Zimmerman emphasizes, "Since releasement lies beyond willing, it stands outside the ordinary distinction between activity and passivity."[7] Releasement is not merely a flight into "mysticism," however, but a movement to restore the boundaries of life: for example, the cosmic rule that life's transitoriness calls for maximizing rather than squandering its opportunities, distributing one's blessings rather than monopolizing them, or assuming modesty rather than pride for one's accomplishments. While the term may at first appear clumsy and even misleading, we can chart the recovery as a transition from "excess" to "economy," where the latter expresses a principle of tending, allocating, preserving over the sphere of one's own benefits. The economy is the "holding" of one's own residence, or house of dwelling, which is not confined to one's material assets but instead includes all the fruits of one's activities, commitments, and relationships.[8] As a case in point, the taking of "inventory" advocated by the Twelve Step Program is not only a mental and moral self-assessment, but also involves examining where one stands in relation to this universal principle or, put another way, the direction one

[6] For a discussion of "biological continuity," and its importance for a phenomenological account of embodiment, see Westling 2014, p. 77. Also see Schalow 2015, pp. 61–82 and Nelson 2004, pp. 65–74.

[7] Zimmerman 1986, p. 241. For an practical application of the concept of mindfulness, as promoting a way of dwelling in the "moment" to combat addictive impulses, see Kiloby 2017, pp. 20–22.

[8] See Schalow 2006, p. 18.

takes along this path of mindfulness. Let us then briefly consider how proceeding along this path shapes the circular movement of the individual's recovery from addiction.

8.1 The Role of Boundaries

As we have seen, everydayness forms the context from which we can discern the human tendency to become addicted, from the individual's vulnerability to the "hook" of desire, to the forces of "peer pressure," not to mention the self-delusional, deceptive tendencies and practices which hasten the "downward spiral" of addiction. Conversely, only at the periphery or "margins" of everydayness are there momentary flashes of opportunity to reverse these self-destructive tendencies if not find the inspiration (or even the "make or break" attitude) to embark upon a path of recovery. Following contemporaries like Karl Jaspers, Heidegger characterized these margins as a "limit" or "boundary" situation, that is, which bring the individual before the precipice of self-decision by marking the connection between life and death, opportunity and risk. Existentially speaking, the early addiction specialists (e.g., Bill Wilson) alluded to something similar when they earmarked the importance of the point of capitulation, of "hitting rock bottom." The question arises, however, whether, short of this (almost near-death experience), a catalyst can induce this transforming experience, i.e., mark the beginning or cornerstone of treatment.

The "rock-bottom" approach harbors an important truth: that the addict him/herself must take the initiative, and that reprimanding, persuading, and cajoling do not work. But that approach also harbors a hazard that instead of being able to pull him/herself out the downward spiral of addiction, the individual will simply plunge further into an inescapable abyss of self-destruction and eventually death. The opposite extreme would take the form of "intervention" or forcibly placing the addict into a rehabilitation program. Yet the caveat here is that success is still predicated on the individual's recognizing the need for help and committing to the lengthy process of recovery. The challenge of intervention is that it must address the basic blind spot of the addict, namely, the stubborn *denial* of his/her addiction or the refusal to recognize his/her problem *as* a problem. Wherever help is sought, there must still be a "catalyst" or an inducement that elicits from the addict the self-initiative to change and commit to a program of recovery.

Let us examine the more familiar setting of "talk-therapy" to find a possible parallel. To state the obvious, even in the case of therapy the patient wants to get better, even if it is painful to confront one's demons. In this setting, the inducement may lie in the most innocuous form, that is, not in what is said or communicated in the therapy, but what remains unsaid and yet a deeper source of communication. As a case in point, let us consider the chief venue of the therapeutic process as rooted in dialogue, that is, between patient and counselor. First, a phenomenological-experiential mode of "openness" prevails in the therapeutic relation as an instance of "being-in-the-world," "being-with-others," and "being-toward-oneself."[9] In this therapeutic context, it is as equally incumbent upon the counselor *not* to impose his/her views and even "theories," as necessary for the patient to be forthcoming in his/her admissions. Accordingly, the dialogue (*Zwiesprache*) is not reducible to the verbal contribution of each participant (either counselor or patient), for the possibility of illumination supersedes both (even the "combined" viewpoints). Instead, to use the term that Heidegger employs in *Contributions to Philosophy*, the "mid-point" (*Wesensmitte*) turns or "swings" the dialogue (and not the other way around), appropriating the "clearing" itself as the space for a resounding silence, that is, as the stillness prefacing all self-discovery. Secondly, the stillness is not only a formal component of the individual's disclosedness, but also *grounds a specific methodological principle within the therapeutic process*: the reservedness of "stepping back" as the directive to govern the counselor's comportment toward the patient.

In simple terms, the direction of the dialogue does not lead to further explication, but instead, swings back the other way, as it were, toward "holding in reserve," withholding, and, ultimately, remaining silent, in order to preserve the mystery of human existence *as* a mystery for the individual to rediscover (for him/herself). The pull of this reservedness, as it were, supersedes even what the counselor *says*, replacing the desire to "help" with a counter invitation and challenge to the patient to search, discover, and embark upon what is his/her "ownmost," in the sojourn of his/her *own* existence. If it is left to the counselor's devices to "figure everything out" for the patient, then the latter can never be moved by the openness of existence and exercise the freedom to pursue his/her unique possibilities. For example, the gradual elevating of the threshold of openness in therapeutic practice enables the patient to experience the deepest sources of sorrow and pain, which, while experienced on one level as an obstacle, can also become a catalyst of self-awakening.[10] Finally, the ownmost dimension of one's individuality is granted only as a "gift." Because it is shrouded in the mystery of silence, the reservedness [*Verhaltenheit*] (of be-ing) [*Seyn*] alone can grant this gift, the miraculous singularity of existence, which can never be awarded as a result of an exclusively human or technical contrivance. The gifting of existence transforms the experience of pain into a resiliency, forbearance, and even wisdom, which cannot be measured in terms of the calculus of the "feel-good" mentality of our technologically driven culture of conspicuous consumption.

In the preceding examples, the possibility of therapeutic practice resides in our situated "dwelling" and hence in the reservedness of silence, which specifically "clears" the arena for this form of human comportment, interaction, and self-discovery.[11] Even the healing of therapeutic practice presupposes the possibility

---

[9] See Schalow 1990–1991, pp. 69–83.

[10] Richardson 2002, pp. 351–354 (esp. p. 352). While counter balancing Heidegger's view of truth as *aletheia*, and Jacques Lacan's psychoanalytic approach to the self, Richardson emphasizes that the disclosure of pain is the beginning of the "real journey."

[11] In section 38 of *Contributions to Philosophy*, Heidegger refers to "*sigetic*" as "reticence in silence," which prefigures the individual's self-disclosedness and provides the hidden dynamic for

of a "breakthrough" in the individual's existence, just as the reservedness of silence grounds whatever is said (in the context of therapy) within the wider expanse of being-in-the-world. Whether or not therapy is explicitly employed as a practice to help the addict toward recovery, the preceding example underscores two important considerations. First, however one gets "there" or whatever path one takes, there must already be a *moment of self-reckoning*. Second, there is a paradox within the recovery process such that the initiative to change is both "closer" than the "within reach" of whatever can be addictive and "farther" than all the well-intended attempts to help by those who "enable" the addict. Because help is simultaneously far and close, there is no "magic bullet" to recovery other than the opportunity that can present itself in the "moment" within the dynamic of human temporality. As is a primary lesson of meditation, the key to changing is to abide within the moment, in order that the potential to change can overturn the reluctance to do so and that possibility can triumph over what circumstances deem to be impossible.

Can we summarize the key elements of addiction and thereby provide a snapshot of the overall process? First, recovery always unfolds against the backdrop of everydayness. The potential to be a self, including the directive of its priorities to be a self, already outline the trajectory of what is possible (and its gradual restricting through the individual's tendency to succumb to his/her indulgences, that is, become addicted). Secondly, recovery is a movement that counter poses or turns against the individual's penchant for dissimulation and deception, and thereby moves in the elliptical arc of temporality in the search for greater self-understanding. Thirdly, in contrast to the inertia of merely reacting to the pull of one's cravings and indulgences, recovery cultivates an awakening to and *responsiveness* toward the (hidden) capability to exercise choice, the "choosing to choose". Through such resolve, the individual extends care and homage to him/herself—prior to caving in to the impulsiveness of his/her desires and the expediency of bending to peer pressure. Fourthly, the "moment" (*Augenblick*) is already ripe with offering the chance for recovery over against the instant of immediate gratification, the playing out of the temporalizing of human existence across the spectrum of owned (authentic) and unowned (inauthentic) existence. The "moment" combines the individual's receptivity to the "gift" of human existence (the infusion of this abundance in and through the self), with its *own* commitment to be "entrusting" thereof (in and through the *response*-ability of choosing and acting). As a temporal-spatial dynamic, recovery entails the *creation of a new life-trajectory* (rather than a stasis of the elimination of negative tendencies and symptoms).

Our phenomenological approach identifies elements that help to form the cornerstone of any recovery process. Though the destination may be the same, the paths are many. Can we map these elements upon the specific procedures that are believed to be necessary for an individual to "work" his/her own "program" recovery?

---

its unfolding. "Reticence in silence stems for the swaying origin of language itself" (1989/1999), p. 80; tr. 55.

8.2 Contrition, Self-Discovery and Disclosedness

There is a simple observation that remains constant throughout this study. The problem of addiction never occurs in a vacuum, but instead arises within a specific historical era, culture, and life-situation (which always involves other people). In Heidegger's terms, being-in-the-world already implies being-with others. Conversely, in the case of any self-discovery, there is an accompanying disclosure of the world and of those who inhabit it. The nexus of these interconnections also applies in the case of the phenomenon of addiction, insofar as it arises within this wider compass of human relationships. In the crucial moment of self-reckoning the individual stands forth in his/her possibility of disclosing and discovering *who* he/she is. The disclosure to myself, and for myself must not end at that point, but must also be for the sake of others, that is, in a way which is enacted in the world–in the social arena, and not just "privately."

Because deceiving oneself and others is a crucial dynamic within addiction, the process of recovery requires reversing these deceptive practices. Because others may also be caught in the web of these deceptive practices, the addict can make amends only by uncovering his/her lies, e.g., cajoling money to buy drugs from a relative under the pretext of a short-term loan in order to pay the electric bill. Indeed, all the machinations and manipulations of others have to be brought out into the open and addressed (*in the "safe-space" of fellowship*) in order to break the impetus to cover up, conceal, and mislead. In the case of sexual addiction, the need to make amends can take a more "personal" form, that is, when one's infidelities have inflicted pain upon one's spouse or partner. In some programs for treating sexual addiction, particularly in the process of couples counseling, a key prerequisite for recovery involves "disclosing" all of one's infidelities or affairs, including names and places.

Heidegger's term "disclosedness" (*Entschlossenheit*) parallels the meaning of the term that addiction specialists use. As a structure of human existence, disclosedness in his sense includes the dynamic that governs the candidness and self-admission within the therapeutic process. That is, to disclose is to bring out into the openness what has previously remain covered up or concealed. For Heidegger, it is important to emphasize that, ontologically speaking, truth as disclosedness always occurs in tension with its opposite.[12] Conversely, the life-drama of the self's struggle with its addictions still mirrors the alternative possibilities of heroism, hope, and transcendence. The condition of openness, and our occupation of a world with others, points to another key element in the recovery process: the discretion of individual choice. For Heidegger, freedom and openness are reciprocal, insofar as the development of one's choices always unfolds within a field of possibilities. Yet there is another aspect to this freedom which often goes unnoticed, and which correlates with the distinctive character of human existence as finite. As Heidegger states in *Being and Time*, "freedom occurs in choosing one possibility, and tolerating not

[12] See Richardson, pp. 353–354. Also see Heidegger 1989/1999, pp. 342–344; tr. 239–241.

having chosen other possibilities."[13] The restriction of the "not" works to the benefit of the individual, however, rather than to the detriment. For it is in the attitude of pursuing each and every source of indulgence, as if the world were one's oyster, that the individual goes awry and falls into the trap of addiction. On the contrary, the discretion of choice, or what Aristotle called "balanced judgment," allows an individual to locate the boundaries that lend measure, proportion, and direction to one's existence. The boundaries, as it were, mirror the uniqueness of my possibility, or, in Heidegger's terms, what is "ownmost". Two implications follow from this observation. First, the measure of uniqueness allows one to discover meaning (and perhaps even joy and happiness) in one's existence, including a sense of gratitude for one's blessings.[14] Second, the more one becomes grounded in his/her individuality, in what is "ownmost," the more one can resist the social forces of conformity, the "peer pressure" to be "cool" and run with the crowd–the mindless going with the flow that creates the opportunity for becoming addicted in the first place and also sustains the addiction(s) afterwards.

In this regard, the starting place of everydayness raises the common question, for addict and non-addict alike, of one's associates and places one ventures, throughout one's sojourn in the world. The encompassing life-situation of each individual already harbors the threefold concern which cuts across simultaneously the need to change the thought patterns and the behavioral patterns of the addict: the "who," the "where," and the "why?". In making better decisions, the recovering addict must not only avoid "using," but also steer away from the very same environmental influences which encourage indulgence, including past associates (poor role models) and popular places (bars). Indeed, sobriety cannot be "won," or even achieved momentarily, by lapsing into the previous habits and returning to the social situations that encourage indulgence. The same places where "social drinkers" congregate become "off limits" to a recovering alcoholic, that is, such a locale occurs "outside the boundaries" of one's field of concern and activities. Analogously, the amusement of a casino for one person may become an overwhelming temptation for someone with a gambling addiction. Of course, when the place of indulgence becomes as *ubiquitous as cyberspace*, the designation of "off-limits" becomes a much more difficult mandate to enforce.

The addict perpetuates the self-delusion that replacing suffering with the transitory euphoria of immediate gratification can serve as a formula to resolve (the *pathos*) of the human predicament. By contrast, the phenomenological appeal to "meaning" as a priority over the search for happiness reveals that the path of recovery (and thereby the transition toward healing, wholeness, health, and fulfillment— if not "salvation" as such) unfolds in the opposite direction. The uncovering of meaning provides the resiliency that makes an encounter with suffering *possible*,

[13] Heidegger 1977b/1962, p. 378; tr. 331.

[14] Frankl shows how the discovery of meaning supersedes happiness as a directive for human existence, by allowing the opposite of joy, or suffering, to supply a catalyst and potential for self-development and growth (beyond the search for immediate gratification). Frankl 1962, pp. 27–27, 106. See Frankl 1980, pp. 105–111.

not as end in itself, but as a catalyst to growth and self-development which curbs (i.e., set limits to the one-sided search for happiness as immediate gratification). Thus, the addict's fast and furious life-style of seeking immediate gratification harbors a unique paradox that the avoidance of suffering exacts an even greater "payment" down the line as a potentially life-transforming *possibility*. In clinical terms, we can invoke the scenario of the addict's "hitting rock bottom." Yet even this description is a metaphor for the need for the exacting of an existential payment in which suffering emerges as the therapeutic counter point to happiness.

Even more fundamentally, the challenge of self-discovery, of uncovering the meaning of one's existence, pre-empts the cultural prejudice in favor of pursing the trappings a life-style of conspicuous consumption. The development of a new life-trajectory requires an ongoing commitment to cultivate and elevate one's capacity for self-awareness. The hope of a new, spiritual awakening does not occur in a vacuum, however, but instead includes the *appropriation of life's lessons*. Without reference to a formal curriculum, we might characterize this kind of "learning," in Søren Kierkegaard's words, as "education in the school of possibility," or, as Hatab states (following Heidegger), becoming educated in one's finitude.[15]

8.3 The Poetics of Transformation

Will scientific and technological advancements ultimately provide a "panacea" for the problems of today? Could the science of the future provide some kind of magical formula to prevent obsessive-compulsive tendencies and virtually "cure" addiction? Given recent breakthroughs in neuroscience and new insights into brain-chemistry,[16] perhaps new models for treating addiction will emerge which more or less bypass the need for spirituality and the Twelve Step Program. In this new age of science, we cannot ignore the promise that mapping the human genome may hold for identifying and even providing the keys to rectify the genetic predispositions for addiction.[17] Yet, as promising as these new advances may be, we are still bound by the limitations of today. Moreover, there remains an ineluctable aspect of human existence which science may not be able to unravel, the simple questions pertaining to "meaning": "who am I?," "why am I here?," or even, more metaphysically, "why is there something rather than nothing?" These perennial questions are as old as civilization itself, and they may continue to be relevant as long as human beings occupy this earth.

These are questions that Martin Heidegger broached throughout his career, albeit in different forms. He also brought a healthy skepticism toward the future promises of science and technology. In the process, he searched for a language that could also

[15] Kierkegaard 1944, pp. 139–145. See Schalow 1989, pp. 160–167. Also see Hatab 2000, p. 127.

[16] See Erickson 2007, pp. 50–73.

[17] See von Stieff 2012, p. 275. Von Stieff points to genetically based "imbalances" in brain chemistry as a "real problem in society" today.

address mystery of existence and could in turn add nuance and depth to the formulation of the most basic philosophical questions. He found in poetry, particularly through his German counterparts Friedrich Hölderlin and Rainer Maria Rilke, as well as in Greek tragedy, a special sounding board from which to elicit new idioms of speech to express what otherwise remains inexpressible in philosophical concepts.[18] Equally as important, Heidegger turned to both art and poetry to arrive at alternative ways of revealing our place in the world, which diverge from the technocratic, one-dimensional, and materialistic world-view of the sciences. Specifically, the creative power of the work of art provides a foremost instance of the opening of a world, thereby distinguishing a unique and singular way of unconcealing (things in their wholeness). Correlatively, by participating and engaging in *poiesis*, human beings are appropriated to this openness, in order that they can relate to all that is (including nature) apart from the technological drive to package everything in a calculative, uniform, and one-dimensional way—that is, in terms of the *gestalt* of the artifice. For poetry, including the highest expression of art, e.g., sculpture, painting, and music, has the power to evoke the most singular of all our experiences, in contrast to the generic formulas of science and technology. Although we cannot go into detail here, Heidegger contrasts the work of art as a form of *technē* (τέχνη) or craftsmanship, that is, as a singularly disclosive process, with a regimented form of technological mass production (including the processing of digital images in cybernetics).

Could poetry in this primordial sense also provide a glimpse into the most hidden, ineluctable, and mysterious powers within the human heart, which foretell any potential for transformation? Could poetry, not only in the novelty of its language, but also its distinctive way of disclosing through the interplay of emotion and thought, inspiration and sublimity, cast light on the confluences of forces that guide the individual's recovery from addiction? Poetic imagery transposes the linear way we ordinarily think into a circle of understanding. Within the movement of this circularity we experience the complementarity of opposites, rather than their polarization. For each point along the circumference of the circle is the same point from another perspective. To take a rather simple example, what the addict first experiences as the greatest adversity can become the greatest opportunity.

The springboard leading to recovery lies precisely in this transforming moment: the breaking down, inverting, and overcoming of the false dichotomies which fuel the fragmented, compartmentalized, deceptive thinking of the addict, i.e., the "either-or," "all or nothing" mentality. We might suggest that this newly founded insight or understanding entails a unique, "existential logic" of its own. Just as traditional logic corrects fallacies or errors in reasoning, so an existential logic offsets the misguided, habituated "rationalizations" that both embody and sustain the addictive personality. Where Twerski emphasizes the need to correct "addictive thinking",[19] I emphasize how an existentially enacted logic (e.g., a self-expressive

[18] See Smith 2013, pp. 126–130, 182.

[19] Twerski 1997, pp. 102–105. For him, "rock bottom" entails a change of perception, in which abstinence is preferable to use. See O'Connor 2016, p. 160. "Each person's 'rock bottom' is the point where she can no longer tolerate the misery.'"

logos) of commitment, candidness, and self-discovery combats the life errors of denial, deception, and dissimulation. Most of all, the *logos* in play reserves a space for the quietude, silence, and reticence by which the individual can first heed a call of self-awakening and start along the path of recovery. The emergence of a new circle of understanding supplants the aforementioned, narrow dichotomies in a movement of integration and discovery, in which opposites are crossed to create a newly centered, balanced and holistic vision of one's place in the world. By analogy, we might say a compass arises with an arrow that always points along the path of recovery, restoring a sense of balance and a sense of purposiveness at each step. Rather than being torn asunder by conflicting views, the recovering addict discovers that the interplay of opposites or pair of contraries point in the same direction (e.g., adversity/opportunity)–only from a different beginning.[20] The fact that the direction of human existence is never given in a straight line speaks not only to its essential finitude, but also points to the circularity of human understanding and its historical situatedness. Let us consider an example.

At the conclusion of his essay, "The Question Concerning Technology," Heidegger cites a provocative line from Hölderlin's poetry: "Near is god and yet hard to find; where danger lies grows salvation also."[21] In marking the zero point between despair and self-recovery, the precipice of destruction also harbors the possibility of a turning around of destiny. A fundamental question arises in Heidegger's thought as to whether technology may harbor an alternative way of revealing ("what is"), that is, in contrast to the emphasis on usefulness, instrumentality, and the narrow focus of our culture of conspicuous consumption. In the mid- 1930s and subsequently, he appealed to the work of art as an example of a disclosedness, which accents the singular vision of creativity over against the generic, instrumentalist claim of usefulness, that is, of how things fit together within a gestalt of machination—the so-called "cog in the machine." For him, *poeisis* epitomizes the creativity by which otherwise hidden possibilities of manifestation can be brought into play and thereby rupture the hegemony of a one-dimensional, technocratic way of revealing. In terms of our study, poetic figures and gestures serve as catalysts to displace and invert the regimentation of an individual's distorted self-images, thereby lighting up possibilities that can speak to the singular concerns distinctive of him or her. The self-mirroring of art, then, can prompt a sense of awakening to an untapped potentiality, of "who I can be." The more humbling that awakening may be, the more easily can the self, by inviting a greater sense of openness, break the shackles of "mechanically" acquiescing to the allure of its indulgences—including their intensification by mediating one's desires through *the rise of artifice and the magnetic pull of cyberspace*. The *technē* of *poeisis* stands in stark contrast to the technological productions or designs of the *artifice*, which infiltrate our daily life from every side and dominate our culture through the media of television and the Internet.

---

[20] See Jung 1974, "The Stages of Life," pp. 94–114.

[21] See Heidegger, 1977a, p. 35.

In our culture of conspicuous consumption, we may easily overlook the "reversal of fortune" which is bound up with our sense of humility or spiritual destination. Precisely where no material article or self-indulgence can fulfill a need, there arises a necessity that turns us toward a "spiritual" plane, that is, to a meaning that cannot be exhausted by the score-keeping of our ordinary calculations of the pluses and minuses of life. In his 1945 essay, appropriately titled "Poverty" ("*Die Armut*"), Heidegger cites one of the most intriguing lines of Hölderlin's poetry: "'For us everything is concentrated on the spiritual, we have become poor in order to become rich.'"[22] Here it would be too facile to develop a comparison between poor and rich as a contrast between material and spiritual well-being. Rather, poverty and wealth are more intimately tied together, insofar as the latter may be found in places we least expect, implying an attitude of attentiveness and forbearance. The ultimate "spirit" of freedom is a releasement toward and openness to possibilities which otherwise may remain hidden and overlooked, but which may also harbor the potential of re-birth in the individual's existence. In the spiritual practice of Meister Eckhart, *Gelassenheit* (releasement) requires a "preparation (*Bereitung*) for receiving a gift, a labor of waiting in spiritual poverty and humility…for his [God's] entrance into the soul's ground…."[23] In a brief passage under the heading "On Letting Go of Things," Heidegger cites this succinct remark from Eckhart: "'Where I will nothing of myself, there wills instead my God.'"[24] Following Eckhart, Heidegger describes this disposition as the resiliency of "letting be," which welcomes, seeks, and appreciates rather than imposes, denies, and dismisses. *Could the "poverty" to which the poet alludes to above suggest a deeper meaning to the self's "bottoming out," that is, to the capitulation which provides the catalyst for the individual to reverse the downward spiral of addiction?*

In the poverty of what one chased with the greatest fervor resides a greater wealth of possibility. *The apparent self-destructiveness of "hitting of rock bottom" reverts into a positive moment, an opportunity for transformation, through the abundance that arises from confronting the impoverishment of one's indulgent cravings.* As Heidegger emphasizes in an earlier discussion, poverty as "being-poor" is not simply a lack or privation, but instead is a basic comportment (*Verhalten*),[25] the endurance of which (e.g., as in the case of self-humbling) sets in motion the steps toward personal transformation (and ultimately releasement). Although this characterization may apply to many of our preoccupations, it perhaps rings most truly in the case of the addict's fixations. For on the other side of enslavement to one's desires lies the emancipation of finding an alternative path of life. In this appeal to this higher spirituality, however, we must also reaffirm the practical measures that help to direct one along this new path. For this path of releasement still requires the priorities that help to sort out the wealth of one's own possibilities from the poverty of

---

[22] See Heidegger 1994/2011, p. 9; tr. 3. Also see Heidegger 2006, p. 378.

[23] Kovacs 1995, p. 44. See Seeburger 1993, pp. 171–174.

[24] Heidegger 2010, p. 103 (and 103n).

[25] Heidegger 1983/1995, pp. 277–278; tr. 195.

satiating the lowest level desires. Indeed, only through the development of these priorities can an individual make sustained progress along the path of life.

Heidegger attributed a "saving power" to *poeisis*, not as an avenue of escape, but as opening forth a new place (*Ort*) or site in which the *self can take back ownership* of its existence and become *open to possibilities other* than those confined to the narrowest technological impulses.[26] By re-*situating* his/her existence within this wider horizon, the individual can cultivate the openness to choose, to reset priorities in ways that make the employment of technology subordinate. The saving power of *poeisis* transposes the self into the openness of its capacity to choose, returning to it the decision to decline or say "no" to the exploitative tendencies of technological society, as much as to affirm or say "yes" to the challenges that technology poses. Technology challenges us to cultivate a new *mindfulness* of our place in the world, within nature, and, indeed, within a global community, without mindlessly identifying with our "ontic" cravings of using and consuming things. To extend the analogy, the path of recovering from addiction requires that the individual to develop a new life-trajectory, which lifts him/her beyond the compulsiveness of his/her desires and restores the freedom to chisel out a new "place" in the world.

In this way, poetics gives a face to the individual struggle, apart from the impersonalizing and pressure to conform to the expectations of our commercialized society; and thereby highlights a new path that the individual can follow in order to rebuff the allure of the commodification of the fetish on the basis of which addiction becomes an epidemic today. Conversely, our study yields a new insight: specifically, the optics through which we can address addiction *as* a problem mirrors those that simultaneously illuminate the limitations, if not the *pitfalls* of technology. Put in Heidegger's terms, the *crisis* of addiction "formally indicates" the complicity with which we embrace each technological advance and thereby reveals the danger of our increasing *dependence* upon them. In this way, *an unusual synergy plays out whereby what is experientially displayed on the micro level of addictive practices also unveils on a macro level the challenge that modern technology poses to us*.

With all its conveniences and comforts, technology lures us to be complacent and to concede the simple fact "the way of least resistance" is usually the best option, thereby drawing upon the deeper tendency to conceal who we are as human beings. But the development of a new life-trajectory reveals the contrary: namely, that the path of taking ownership for one's existence requires the greatest resiliency in enduring, transforming, and recovering from the deepest sorrows and pain. In today's "feel-good" culture, we cannot overlook the fact that as embodied beings we are vulnerable to the larger *pathos* of the human predicament. In the encounter with melancholy there is also the promise of joy, in hardship the prospect of transcendence. In *Dunio Elegies*, Rilke describes the withdrawn depths (*Ab-grund*) of our existence as the "Pearls of Pain and the fine-spun Veils of Patience" (*Perlen des Leids und die feinen Schleier der Duldung*).[27] As such, our moods and affects are also windows in the larger expanse of possibilities that come to light in the *turning*

[26] For further discussion of the saving power of art, see Radloff 2007, pp. 318–321.

[27] Rilke 1963, p. 81.

point of the self's discovery of its new life-trajectory. This turning point is nothing else than the anticipatory-awakening of the moment of *self-transcendence*, in which we become open to experience the abundance of all that life has to offer even in the simplest blessings. The gifting of human existence continues to extend that turning point, which harbors the temporal-spatial dynamic of transformation, even in the breach of the self's hitting rock bottom.

The wider-spread, self-destructive character of addiction not only exposes the danger of technology, but also highlights the challenge that its global impact has upon us to reexamine our place on the earth and the deeper mystery of being. The crucible of conflict through which individuals experience their capacity to change mirrors the incubation period (*Incubationszeit*) by which humanity becomes awakened to its deeper capacity to "let be," and abide within a greater harmony despite the drive toward technological domination. The "turning" (*die Kehre*) which Heidegger envisioned still requires human engagement by enduring the lengthy period of a "counter turning" (*Widerkehre*) and *standing in the breach of the conflict of which our embodied selves provide the locus.*[28] In this period, the hardship of strife (e.g., forbearance) reverts into the abundance of receiving the "gift" (*Es gibt*) of our existence. By the same token, the crisis we experience today as the downside of technology—of which the epidemic of addiction is an instance—may also "turn around" into an opportunity to break the shackles of our dependency and discover a deeper source of freedom beyond the relentless drive for power and the acquisition of material goods.

In this regard, the falling-prey that the individual experiences as the existential ground of addiction, of its fetish-like craving, may extend even more deeply in the fragility of the potential fault line or fissure (*Zerklüftung*), the unstable terrain or ground (*Ab-grund*), on which our technological society rests. Quoting the poet, Hölderlin, Heidegger emphasized that: "'We are a sign that is not yet read.'"[29] Whether such signs appear on a personal front, as in the case of addiction, or an environmental front, as in the case of global warming,[30] the concerns that are thereby evoked challenge us to address ourselves through another lens than that of the relentless drive for material progress. Could the *crisis of addiction,* then, point us to the juncture where the destructive forces of technology also awaken us to its hidden challenges? Questions like these call for an integrative approach that only the philosophical thrust of phenomenology can provide, by mapping them back upon the larger life-drama of what it means to be human.

---

[28] Heidegger 1989/1999, p. 407; tr. 287. "Turning is counter-turning [*Widerkehre*]…."

[29] Heidegger 1968, pp.10–11.

[30] See Irwin 2008, pp. 29–30.

8.4 Conclusion

In the haste to find a treatment and even a "cure" for addiction, we fail to consider what this phenomenon has to tell us about the frailties of human life, and how our triumphs are always mired in the conflicts we overcome. When viewed within a larger historical context, addiction is not just about the plight of the addict, but instead is also a signpost in the search for a spiritual guidance that is otherwise lacking in our culture of conspicuous consumption. The relentless drive of technological progress does not alleviate the poignancy of the basic questions concerning the human predicament, but, when viewed through the lens of addiction, increases their overall relevance. Indeed, these questions acquire renewed *importance when expressed through the unique life-trajectory of each individual's sojourn on the earth*. The humility of affirming the gift of existence, the abundance of life's possibilities, provides the leaping-off point for the edifying movement of self-transcendence, the horizon-opening development of one's potentiality to be.

In this regard, the mission of a phenomenological study of addiction is to cultivate a wider perspective than can be found in a clinical dissection of various forms of addictive behavior. We can have discovered that the phenomenon of addiction, far from defining a source of pathology, serves as a prism to illuminate a broad spectrum of issues pertaining to the human condition and the cultural *ethos* of our technological society. Conversely, the challenge of developing a phenomenology of addiction requires a sustained and concentrated effort to combine philosophical and practical horizons, the fusion of which promises new insights about our humanity and the historical path on which our civilization is traveling.

References

Aho, Kevin. 2009. *Heidegger's neglect of the body*. Albany: SUNY Press.

Frankl, Victor. 1962. *Man's search for meaning*. Boston: Beacon Press.

———. 1980. *The doctor & the soul: From psychotherapy to logotherapy*. New York: Random House.

Hatab, Lawrence J. 2000. *Ethics and finitude. Contributions to a Heideggerian moral philosophy*. Lanham: Rowman & Littlefield.

Heidegger, Martin. 1968. *What is called thinking?*, trans. J. Glenn Gray. New York: Harper & Row.

Heidegger, Martin. 1962. *Being and time*, trans. John Macquarie and Edward Robinson. New York: Harper & Row.

———. 1977a. *The question concerning technology and other essays*, trans. William Lovitt. New York: Harper & Row.

———. 1977b. *Sein und Zeit, GA 2*. Vittorio Klostermann: Frankfurt am Main.

———. 1983. *Die grundbegriffe der metaphysik*, GA 29/30. Frankfurt am Main: Vittorio Klostermann.

———. 1989. *Beiträge zur philosophie (Vom Ereignis)*. Frankfurt am Main: Vittorio Klostermann.

———. 1994. Die Armut. *Heidegger Studies* 9: 9–14.

———. 1995. *The fundamental concepts of metaphysics*, trans. Nicholas Walker and William McNeill. Bloomington: Indiana University Press.

———. 1999. *Contributions to philosophy (From Enowning)*, trans. Parvis Emad and Kenneth Maly. Bloomington: Indiana University Press.

———. 2001. *Zollikon seminars*, ed. Medard Boss, trans. F. Mayr and R. Askay. Evanston: Northwestern University Press.

———. 2006. *Mindfulness*. Translated by Parvis Emad and Kenneth Maly. London: Continuum.

———. 2010. *Country path conversations*, trans. Bret W. Davis. Bloomington: Indiana University Press.

———. 2011. Poverty, trans. T. Kalary and F. Schalow. In *Heidegger, translation, and the task of thinking: Essays in honor of Parvis Emad*, ed. F. Schalow. Dordrecht: Springer.

Irwin, Ruth. 2008. *Heidegger, politics and climate change: Risking it all*. London: Continuum.

Jung, C. G. 1974. *Modern man in search of a soul*, trans. W. S. Dell and Cary F. Baynes. New York: Harcourt, Brace, Jovanovich.

Kierkegaard, Søren. 1944. *The concept of dread*, trans. Walter Lowrie. Princeton: Princeton University Press.

Kiloby, Scott. 2017. *Natural rest for addiction. A radical approach to recovery through mindfulness and awareness*. Oakland: Non-Duality Press.

Kovacs, George. 1995. The way to ultimate meaning in Meister Eckhart. *Ultimate Meaning and Reality* 18 (1): 34–53.

Kurzweil, Ray. 1999. *The age of spiritual machines*. New York: Viking Penguin.

O'Connor, Peg. 2016. *Life on the rocks: Finding meaning in addiction and recovery*. Las Vegas: Central Recovery Press.

Radloff, Bernhad. 2007. *Heidegger and the question of national socialism: Disclosure and gestalt*. Tornonto: University of Toronto Press.

Richardson, William J. 2002. Psychoanalytic praxis and the truth of pain. In *Heidegger and practical philosophy*, ed. F. Raffoul and D. Pettigrew. Albany: SUNY Press.

Rilke, Rainer Maria 1963. *Dunio elegies*, trans. J. B. Leishman and Stephen Spender. New York: Norton & Norton.

Schalow, Frank. 1989. Dread in a post-existentialist era: Kierkegaard re-considered. *Heythrop Journal* 23 (2): 160–167.

———. 2006. *The incarnality of being: The earth, animals, and the body in Heidegger's thought*. Albany: SUNY Press.

———. 2015. Heidegger, the law of being, and animal protection laws. *Ethics & the Environment* 20 (2): 61–82.

Seeburger, Francis. 1993. *Addiction and responsibility*. New York: Crossroads.

Smith, David Nowell. 2013. *Sounding/silence: Martin Heidegger and the limits of poetics*. Bronx: Fordham University Press.

Twerski, Abraham. 1997. *Addictive thinking*. Center City: Hazeldon.

von Stieff, Fredrick. 2012. *Brain in balance: Understanding the genetic and neurochemistry behind addiction and sobriety*. Tucson: Ghost Images Publishing.

Westling, Louise. 2014. *The logos of the living world. Merleau-Ponty, animals, and language*. Bronx: Fordham University Press.

Zimmerman, Michael E. 1986. *Eclipse of the self*. Athens: Ohio University Press. (revised).

References

Agamben, Giorgio. 1999. *Potentialities*, ed and trans. Daniel Heller-Roazen. Stanford: Stanford University Press.
Aho, Kevin. 2007. Gender and time: Revisiting the question of Dasein's neutrality. *Epoché* 12 (1): 137–155.
———. 2009. *Heidegger's neglect of the body*. Albany: SUNY Press.
Aho, James, and Kevin Aho. 2008. *Body matters. A phenomenology of sickness, disease, and illness*. Lanham: Lexington Books.
Alexander, Bruce K. 2010. *Globalisation of addiction: A study in the poverty of spirit*. Oxford: Oxford University Press.
Aristotle. 1998. *Nicomachean Ethics*, trans. David Ross. Oxford: Oxford University Press.
Armstrong, William. 1994. Cyberspace and the relation between being and place. *Southwest Philosophy Review* 10 (2): 33–47.
Becker, Ernst. 1973. *The denial of death*. New York: The Free Press.
Bill, W. 1953. *Twelve steps and twelve traditions*. New York: AA Publishing.
Binswanger, Ludwig. 1958. The existential analysis school of thought, trans. Ernst Angel. In *A New dimension in psychology & psychiatry*, ed Rolo May. New York: Basic Books.
Boss, Medard. 1949. *Meaning and content of sexual perversions: A daseinanalytic approach to the psychopathology of the phenomenon of love*, trans. Liese Lewis Abell. New York: Grune & Stratton.
———. 1963. *Psychoanalysis and Daseinanalysis*, trans. Ludwig Lefebre. New York: Basic Books.
Bultmann, Rudolf. 1959. *Jesus Christ & mythology*. New York: Scribners.
———. 1961. New Testament and mythology. In *Kerygma and Myth: A Theological Debate*, ed. Hans Werner Bartsch, trans. Reginald H Fuller. New York: Harper & Row.
Capobianco, Richard. 1993. Heidegger and Jung: Dwelling near the source. *Review of Existential Psychology & Psychiatry* 21 (1): 50–59.
Carnes, Patrick. 1997. *The betrayal bond*. Deerfield Beach: Health Communications.
Cicoan, C. 2008. The question of the living body in Heidegger's analytic of dasein. *Research in Phenomenology* 38: 72–89.
Kovacs, George. 2015. *Thinking and be-ing in Heidegger's Beiträge zur philosophie (Vom Ereignis)*. Bucharest: Zeta Books.
Crowell, Stephen. 2013. *Normativity and phenomenology in Husserl and Heidegger*. Cambridge: Cambridge University Press.

F. Schalow, *Toward a Phenomenology of Addiction: Embodiment, Technology, Transcendence*, Contributions To Phenomenology 93,
DOI 10.1007/978-3-319-66942-7

de Gennaro, Ivo. 2012. Nietzsche: Value and the economy of the will to power. In *Sources and readings on a key concept of the globalized world*, ed. Ivo de Gennaro, 201–233. Brill: Leiden & Boston.

de Spinoza, Benedict. 1961. Ethics. In *The rationalists*. New York: Random House.

Debord, Guy. 1995. *Society of the spectacle,* trans. Donald Nicholson-Smith. New York: Zone Books.

Derrida, Jacques. 2001. *Geschlecht*: Sexual difference and ontological difference. In *Feminist interpretations of Martin Heidegger*, ed. Nancy J. Holland and Patricia Huntington, 53–72. University Park: Penn State University Press.

Dick, B. 1992. *Anne Smith's journal (1933–1939)*. Maui: Paradise Research Publications.

———. 1998. *The oxford group & alcoholic anonymous*. Maui: Paradise Research Publications.

Dillon, M.C. 1993. Sex, time, and love: Erotic temporality. In *Sex, love, and friendship: Studies of the society for the philosophy of sex and love 1977–1992*, ed. Alan Soble, 316–325. Amsterdam: Rodopi.

Dostoevsky, Fyodor. 1968. The gambler, trans. Constance Garnett. In *Great short works of Fyodor Dostoevsky*. New York: Harper & Row.

Drummond, Henry. 1885. *Natural Law in the Spiritual World*. London: Hodden & Stoughton.

Emad, Parvis. 1985. Boredom as limit and disposition. *Heidegger Studies* 1: 63–78.

———. 2013. *"History" and "nothingness" in Heidegger and Nietzsche: Learning from Beiträge*. Budapest: Societas Philosophia Classica.

Emad, Parvis, and Kenneth Maly. 1999. Translators' foreword to Martin Heidegger, *Contributions to philosophy (From Enowning)*. Bloomington: Indiana University Press.

Erickson, Carlton. 2007. *The science of addiction: From neurobiology to treatment*. New York: W.W. Norton.

Foucault, Michel. 1973. *Madness and civilization: A history of insanity in the age of reason*, trans. Richard Howard. New York: Random House.

Frankl, Viktor. 1962. *Man's search for meaning*. Boston: Beacon Press.

———. 1980. *The doctor and the soul: From psychotherapy to logotherapy*, trans. Richard Winton and Clara Winton. New York: Random House.

Freud, Sigmund. 1959. *Future of an illusion,* trans. James Strachey. New York: W. W. Norton.

Gadamer, Hans-Georg. 1967. On the scope and function of hermeneutical reflection, trans. G. B. Hess and Richard Palmer. In *Philosophical hermeneutics,* ed. David Linge. Berkeley: The University of California Press.

Gendlin, Eugene T. 1987. A philosophical critique of narcissism. In *Pathologies of the modern self*, ed. David Michael Levin. New York, New York University.

Gnuse, Robert. 1989. *Heilsgeschichte as a model for biblical theology*. Lanham: University Press of America.

Groth, Miles. 1990/1991. Human being and existence: The beginnings of an existential psychology. *Review of Existential Psychology & Psychiatry* 22: 116–139.

———. 2001. Existential psychotherapy today. *Review of Existential Psychology & Psychiatry* 25: 7–27.

Guignon, Charles. 2000. Philosophy and authenticity: Heidegger's search for a ground for philosophizing. In *Heidegger, authenticity, and modernity: Essays in honor of Hubert L. Dreyfus*, ed. Mark Wrathall and Jeff Malpas, vol. 1, 79–100. Cambridge: The MIT Press.

Hannush, M.J. 2012. Review of world, affectivity, and trauma: Heidegger and Post-Cartesian psychology. *Journal of Phenomenological Psychology* 42 (2): 217–221.

Harrington, Michael. 1967. *The twilight of capitalism*. New York: Simon & Schuster.

Hatab, Lawrence J. 2000. *Ethics and finitude: Heideggerian contributions to moral philosophy*. Lanham: Rowman & Littlefield.

———. 2016. The hurdle of words: Language, being, and philosophy in Heidegger. In *Hermeneutical Heidegger*, ed. Michael Rowler and Ingo Farin, 262–282. Evanston: Northwestern University Press.

Hegel, G.W.F. 1974. *On art, religion, philosophy: Introductory lectures to the realm of absolute spirit*, trans. J. Glenn Gray. New York: Harper & Row.

———. 1976. *The phenomenology of spirit*, trans. A. V. Miller. Oxford: Oxford University Press.

Heidegger, Martin. 1962. *Being and time*, trans. John Macquarrie and Edward Robinson. New York: Harper & Row.

———. 1968. *What is called thinking?* trans. J. Glenn Gray. New York: Harper & Row, Publishers.

———. 1976. *Was ist Metaphysik? In Wegmarken, GA 9*. Vittorio Klosterman: Frankfurt am Main.

———. 1977a. *The question concerning technology and other essays*, trans. William Lovitt. New York: Harper & Row.

———. 1977b. *Sein und Zeit*, GA 2. Frankfurt am Main: Vittorio Klostermann.

———. 1978. *Metaphysische anfansgründe der logik im ausgang von Leibniz*, GA 26. Frankfurt am Main: Vittorio Klostermann.

———. 1982. *Parmenides*, GA 54. Frankfurt am Main: Vittorio Klostermann.

———. 1983. *Die grundbegriffe der metaphysik*, GA 29/30. Frankfurt am Main: Vittorio Klostermann.

———. 1984. *The metaphysical foundations of logic*, trans. Michael Heim. Bloomington: Indiana University Press.

———. 1985. *Schelling's treatise on the essence of human freedom*. Athens: Ohio University Press.

———. 1988. *Ontology (Hermeneutik der faktizität)*, GA 63. Frankfurt am Main: Vittorio Klostermann.

———. 1989. *Beiträge zur philosophie (Vom Ereignis)*, GA 65. Frankfurt am Main: Vittorio Klostermann.

———. 1991. *Kant und das problem der metaphysics*, GA 3. Frankfurt am Main: Vittorio Klostermann.

———. 1992. *Parmenides*, trans. A. Schuwer and R. Rojcewicz. Bloomington: Indiana University Press.

———. 1994. Die Armut. *Heidegger Studies* 9: 9–14.

———. 1995. *The fundamental concepts of metaphysics*, trans. Nicholas Walker and William McNeill. Bloomington: Indiana University Press.

———. 1997. *Kant and the problem of metaphysics*, trans. Richard Taft. Bloomington: Indiana University Press.

———. 1998. *Ontology—The hermeneutics of facticity*, trans. John van Buren. Bloomington: Indiana University Press.

———. 1999. *Contributions to philosophy (From Enowning)*. Bloomington: Indiana University Press.

———. 2000a. *Elucidations of Hölderlin's poetry*, trans. Keith Hoeller. Amherst: Humanity Press.

———. 2000b. Die frage nach der technik and Überwindung der metaphysik. In *Vorträge und aufsätze*, GA 7. Frankfurt am Main: Vittorio Klostermann.

———. 2000b. *Vorträge und aufsätze*, GA 7. Frankfurt am Main: Vittorio Klostermann.

———. 2001. *Zollikon seminars*, ed. Medard Boss, and trans. F. Mayr and R. Askay. Evanston: Northwestern University Press.

———. 2003. *Four seminars*, trans. A. Mitchell and F. Raffoul. Bloomington: Indiana University Press.

———. 2006. *Mindfulness*, trans. Parvis Emad and Kenneth Maly. London: Continuum.

———. 2010. *Country path conversations*, trans. Bret W. Davis. Bloomington: Indiana University Press.

———. 2011. "Poverty." trans. F. Schalow and T. Kalary. In *Heidegger, translation, and the task of thinking: Essays in honor of Parvis Emad*, ed. F. Schalow. Dordrecht: Springer.

———. 2012. *Bremen and Freiburg lectures: Insight into that which is and Basic principles of thinking*, trans. Andrew J. Mitchell. Bloomington: Indiana University Press.

Huxley, Aldous. 1954. *The doors of perception*. New York: Harper & Row.

———. 1968. *Brave new world*. New York: Harper & Row.

Ihde, Don. 2010. *Heidegger's technologies: Postphenomenological perspectives*. Bronx: Fordham University Press.
Irwin, Ruth. 2008. *Heidegger, politics and climate change: Risking it all*. London: Continuum.
Jonas, Hans. 2001. *The gnostic religion*. Boston: Beacon Press.
Jung, C.G. 1957. *The undiscovered self*. New York: New American Library.
———. 1958. *Psychology and religion: West and east*, trans. R. F. C. Hull. Princeton: Princeton University Press.
———. 1965. *Memories, dreams, reflections*, trans. Richard and Clara Winston. New York: Random House.
———. 1969. *The structure and dynamics of the psyche*, trans. R. F. C. Hull. Princeton: Princeton University Press.
———. 1971. Individual dream symbolism in relation to alchemy, trans. R. F. C. Hull. In *The portable Jung*, ed. Joseph Campbell, 323–455.
———. 1974. *Modern man in search of a soul*. trans. by W. S. Dell and Cary F. Baynes. New York: Harcourt, Brace, Jovanovich.
Kant, Immanuel. 1965. *Critique of pure reason*, trans. Norman Kemp Smith. New York: St. Martin's Press.
———. 1981. *Grounding for the metaphysics of morals*, trans. James W. Ellington. Indianapolis: Hackett Publishing.
———. 1996. *The metaphysics of morals*, trans. Mary Gregor. Cambridge: Cambridge University Press.
———. *Critique of judgment*, trans J. H. Bernhard. New York: Macmillan Publishers.
Kardaras, Nicholas. 2016. *Glow kids. How screen addiction is hijacking our kids and how to break the trance*. New York: St. Martin's Press.
Kemp, Ryan. 2009. The temporal dimension of addiction. *Journal of Phenomenological Psychology* 40 (1): 1–18.
Khaleghi, Morteza, and Karen Khaleghi, 2011 *The anatomy of addiction: Overcoming the trigger that stands in the way of recovery*. St. Martins' Press.
Kierkegaard, Søren. 1944. *The concept of dread*, trans. Walter Lowrie. Princeton: Princeton University Press.
———. *Philosophical fragments*, trans. Walter Lowrie. Princeton: Princeton University Press.
Kiloby, Scott. 2017. *Natural rest for addiction. A radical approach to recovery through mindfulness and awareness*. Oakland: Non-Duality Press.
Kirkham, Nin. 2016. Recognizing our place in the world. *Environmental Ethics* 38 (1): 98–119.
Kovacs, George. 1982a. Phenomenology and logotherapy. In: *Analytica Frankliana: The Proceedings of the First Congress of Logotheraphy*, ed. Sandra Wawrytko, 33–45.
———. 1982b–1983. The personalistic understanding of the body and sexuality in Merleau-Ponty. *Review of Existential Psychology and Psychiatry* 18: 207–217.
———. 1986. Phenomenology of work and self-transcendence. *Journal of Value Inquiry* 20: 195–207.
———. 1990. The way to God as absolute transcendence. *American Catholic Philosophical Quarterly* 64: 240–249.
———. 1995. The way to ultimate meaning in Meister Eckhart. *Ultimate Reality and Meaning* 18 (1): 34–53.
———. 2013. Skepticism and logos in Frankl and Nietzsche. *Existentia* 23 (3–4): 309–324.
———. 2015. *Thinking and be-ing in Heidegger's Beiträge zur Philosophie (Vom Ereignis)*. Bucharest: Zeta Books.
Kovel, Joel. 1987. Schizophrenic being and technocratic society. In *Pathologies of the self*. Edited with an Introduction by Michael David Levin, 330–348. New York: New York University Press.
Kurtz, Ernst. 1991. *Not-God. A history of alcoholic anonymous*. Center City: Hazeldon.
Kurzweil, Ray. 1999. *The age of spiritual machines*. New York: Viking Penguin.
Langlais, Michael. 2003. *The truth of love: Heidegger, Jung, and the idea of transformation*. Union Institute and University [Doctoral Dissertation].

Lawford, Peter Kennedy. 2014. *What addicts know: 10 lessons from recovery to benefit everyone*. Foreword by Drew Pinsky. Dallas: BenBella Books, Inc.
Lawler, Leonard. 1992. *Imagination and chance*. Albany: SUNY Press.
Lefebvre, Henri. 1968. *The sociology of Marx*. New York: Random House.
Levin, David Michael, ed. 1987. *Pathologies of the modern self: Postmodern studies of narcissism, schizophrenia, and depression*. New York: New York University Press.
Marcuse, Herbert. 1955. *Eros and civilization*. New York: Random House.
Marx, Karl. 1963. In *The Marx-Engels reader*, ed. Robert C. Tucker. New York: W. W. Norton & Company.
Mcgrath, A. 1985. *Luther's theology of the cross*. New York: Blackwell.
Merleau-Ponty, Maurice. 1961. *Phenomenology of perception*, trans. Colin Smith. London: Routledge.
Mitchell, Jane Valez, and Sandra Mohr. 2011. *Addict nation: An intervention of America*. Deerfield: Health Communications, Inc.
Morris, David. 2001. Lived time and absolute knowing: Habit and addiction from "Infinite Jest to the *Phenomenology of Spirit*". *Clio* 30(4): 409–412.
Nakken, Craig. 1996. *The addictive personality*. Center City: Hazelden.
Nelson, Eric Sean. 2004. Responding to heaven and earth. Daoism, Heidegger, and the earth. *Environmental Philosophy* 1 (2): 65–74.
O'Connor, Peg. 2016. *Life on the rocks. Finding meaning in addiction and recovery*. Las Vegas: Central Recovery Press.
Overgaard, S. 2004. Heidegger on embodiment. *Journal of the British Society for Phenomenology* 35 (2): 116–131.
Pearl, Joel. 2013. *A question of time: Freud in light of Heidegger's temporality*. Amsterdam: Rodopi.
Peck, M. Scott. 1979. *The road less traveled*. New York: Simon & Schuster.
Pinsky, Drew. 2003. *Cracked: Putting broken lives together again*. New York: Harper-Collins.
Polt, Richard. 2015. A Heideggerian critique of cyberbeing. In *Horizons of authenticity in phenomenology, existentialism, and moral psychology: Essays in honor of Charles Guignon*, 179–197. Dordrecht: Springer.
Quinones, Sam. 2015. *Dreamland: The true story of America's opiate epidemic*. New York: Bloomsbury.
Radloff, Bernhard. 2007. *Heidegger and the question of national socialism: Disclosure and gestalt*. Toronto: The University of Toronto Press.
———. 2014. Contra antiphon the sophist: Aristotle, Heidegger and the planetary order. *Existentia* 24: 283–322.
Richardson, J., and S.J. William. 1988. The place of the unconscious in Heidegger. In *Heidegger and psychology*, ed. Keith Hoeller, 187–194. Atlantic Highlands: Humanities Press.
———. 2002. Psychoanalytic praxis and the truth of pain. In *Heidegger and practical philosophy*, ed. F. Raffoul and D. Pettigrew, 339–358. Albany: SUNY Press.
Ricoeur, Paul. 1978. The critique of religion, trans. R. Bradley DeFord,. In *The Philosophy of Paul Ricoeur: An Anthology of his Work*. Boston: Beacon Press.
Rilke, Rainer Maria. 1963. *Dunio elegies*, trans. J. B. Leishman and Stephen Spender. New York: Norton & Company.
Risser, James. 2012. *The life of understanding: A contemporary hermeneutics*. Bloomington: Indiana University Press.
Romanyshyn, Robert B., and Brian J. Whalen, 1987. Depression and the American dream: The struggle with home. In *Pathologies of the modern self*. Edited, with an Introduction, by David Michael Levin. New York: New York University Press.
Schalow, Frank. 1985. The hermeneutical design of Heidegger's analysis of guilt. *Southern Journal of Philosophy* 23: 361–376.
———. 1989. Dread in a post-existentialist era: Kierkegaard re-considered. *Heythrop Journal* 23 (2): 160–167.

———. 1991/1992. Guilt and the unconscious. The critical distance between phenomenology and psychoanalysis. *Review of existential psychology and psychiatry* 22: 69–83.
———. 1993. Willing as the genuine postscript of modern thought: At the crossroads of an anomaly. *Epoché* 3: 77–104.
———. 2000. Heidegger and the question of economics. *American Catholic Philosophical Quarterly* 74 (2): 249–267.
———. 2001a. Eros and the God-question: At the crossroads between hermeneutics and psychoanalysis. *Review of existential psychology & psychiatry* 25: 79–91.
———. 2001b. *Heidegger and the quest for the sacred: From thought to the sanctuary of faith.* Dordrecht: Kluwer Academic Publishers.
———. 2003. Everydayness and the question of human addiction: A phenomenological approach. *Southwest Philosophy Review* 19 (2): 17–30.
———. 2006. *The incarnality of being: The earth, animals, and the body in Heidegger's thought.* Albany: SUNY Press.
———. 2013. *Departures: At the crossroads between Heidegger and Kant*. Berlin: de Gruyter.
———. 2015. Heidegger, the law of being, and animal protection laws. *Ethics & the Environment,* 20/2 (Fall): 6182.
———. 2016. A Diltheyan loop? The methodological side of Heidegger's interpretation of Kant. *Frontiers of Philosophy in China,* 11/3. ed. Eric Sean Nelson, 377–394.
———. 2017. Hume and Kant on imagination: Thematic and methodological differences. In *Kant and the Scottish Enlightenment*, ed. Elizabeth Robinson and Chris W. Surprenant, 213–229. London: Routledge.
Schalow, Frank, and Alfred Denker. 2010. *Historical dictionary of Heidegger's philosophy.* Lanham: Scarecrow Press.
Schelling, F.W.J. 1994. *Idealism and the endgame of theory: Three essays,* ed. and trans. Thomas Pfau. Albany: SUNY Press.
Scott, Charles E. Heidegger and psychoanalysis. *Heidegger Studies*, 6: 131–141.
Seageller, Stephen, and M. Berger. 1989. *The wisdom of the dream.* Boston: Shambhala.
Seeburger, Francis. 1993. *Addiction and responsibility: An inquiry into the addictive mind.* New York: Crossroads.
Smith, David Nowell. 2013. *Sounding/silence: Martin Heidegger at the limits of poetics*. Fordham University Press.
Steeves, James B. 2004. *Imagining bodies: Merleau-Ponty's philosophy of imagination*. Pittsburgh: Duquesne University Press.
Steiner, Claude. 1971. *Games alcoholics lay*. New York: Grove Press, Inc..
Stieff, Fredrick von M.D. 2012. *Brain in balance: Understanding the genetic and neurochemistry behind addiction and sobriety*. Tucson: Ghost Images Publishing.
Stolorow, Robert D. 2011. *World, affectivity, and trauma: Heidegger and Post-Cartesian Psychology*. London: Routledge.
———. Heidegger and post-Cartesian psychoanalysis. In *The Bloomsbury Companion to Heidegger*, ed. F. Raffoul and E. Nelson, 451–458. London: Bloombury.
Summers, Frank. 2013. *The psychoanalytic vision: The experiencing subject, transcendence, and the therapeutic Process*. London: Routledge.
Twerski, Abraham J. 1997. *Addictive thinking*. Center City: Hazelden.
Valverde, M. 1998. *Diseases of the will: Alcohol and the dilemmas of freedom.* Cambridge: Cambridge University Press.
van Buren, John. 1992. Martin Heidegger, Martin Luther. In *Reading Heidegger from the start: Essays in his earliest thought*, ed. Theodore Kisiel and John van Buren, 159–172. Albany: SUNY Press.
———. 1994. *The young Heidegger: Rumor of the hidden king*. Bloomington: Indiana University Press.
———. 2016. Aristotle Luther Heidegger: Travelling a forgotten path of thought. *Existentia* 26 (3–4): 257–273.

Westling, Louise. 2014. *The logos of the living world: Merleau-Ponty, animals, and language*. Bronx: Fordham University Press.

Wilshire, Bruce. 2003. *Wild hunger: The primal roots of modern addiction*. Lanham: Roman and Littlefield.

Yeshe, Lama. 1987. Introduction to tantra. In *A vision of totality*. Boston: Wisdom Publication.

Zimmerman, Michael E. 1986. *Eclipse of the self: The development of Heidegger's concept of authenticity*. Athens: Ohio University Press (revised).

———. 1990. *Heidegger's confrontation with modernity: Technology, politics, art*. Bloomington: Indiana University Press.

———. 1995. Ontical craving versus ontological desire. In *From phenomenology to thought, errancy, and desire: Essays in honor of William J. Richardson*, ed. Babette Babich. Dordrecht: Kluwer Academic Publishers

Index

F. Schalow, *Toward a Phenomenology of Addiction: Embodiment, Technology, Transcendence*, Contributions To Phenomenology 93,
DOI 10.1007/978-3-319-66942-7

www.ingramcontent.com/pod-product-compliance
Lightning Source LLC
Chambersburg PA
CBHW070204170425
25202CB00015B/63
* 9 7 8 3 3 1 9 6 6 9 4 1 0 *